Copyright © 2

All rights reserved

No part of this book may be reproduced, or stored in a retrieval system, or transmitted in any form or by any means, electronic, mechanical, photocopying, recording, or otherwise, without express written permission of the publisher.

Printed in the United States of America

For Bruce Donaldson.
"Avoid large places at night. Keep to the Small"
-from The Boy Who Drew Cats

CONTENTS

Copyright

Dedication

Chapter 1: A Crash Course in the Process of Intelligence   2

Chapter 2. Tools of the SIGINT Trade   27

Chapter 3. SIGINT Antennas   64

Chapter 4. Signals Intelligence Collection and Targeting   88

Chapter 5. Tactical Signals Exploitation   105

Appendix A: Exploiting Individual Sources of Communications   115

APPENDIX B: SIGNALS INTELLIGENCE TASKING CHART (F3EAD)   130

CHAPTER 1: A CRASH COURSE IN THE PROCESS OF INTELLIGENCE

"We kill people based on metadata" -
General Michael Hayden, Director, National Security Agency

The battlefield of today and into the future is a complex animal, defined through internecine conflict as much as it it between national borders. Now well into the Fourth Generation of War as William S. Lind would describe, there are rarely those neat, clean lines of transition or fronts. The contemporary conflict in Ukraine aside, the battlefield is as much rural fields to urban streets or the cyber domain, fought between uniformed actors and civilians alike all sharing the roles of combat. Intelligence is weaponized information driving that fight. It is changing the face of what we understand and recognize as Unconventional Warfare and with it, those evolving requirements of Intelligence. This book specifically deals with communications and specifically the signals intelligence domain, in an effort to simultaneously instruct the methods by which the small unit can utilize these targeting methods with common, off the shelf equipment while recognizing what threats are presented by Radio Frequency (RF) emissions. Communications, both conscious and unconscious, must be viewed as a double edged sword which cuts both ways. It is as much a necessary enabler as it is one which has very real, and very deadly, consequences. The intelligence itself is, in fact, the weapon by which the tools of exploitation, be it a Sniper Observer team waiting in ambush, an electronic attack, a drone strike, or anything in between becomes the dramatic close to the hunt.

The blurry lines between conventional and unconventional warfare have illustrated the ever more important requirement of both parties in a conflict to place a serious emphasis on signals intelligence at every level. In the context of a conventional war, as we are currently bearing witness in Ukraine, the mapping of orders of battle both physical and electronic serve to recognize a chain of command, its capabilities, and the very movement orders themselves. Both parties walk a fine line, seeking to simultaneously exploit the enemy's pitfalls while avoiding their own.

Unconventional Warfare is very much the same, simple, yet maddeningly complex. The very nature of unconventional warfare since the Cold War has found itself in a constant state of evolution not in techniques but rather in terms of capabilities offered to its practitioners. Populist movements, seeking to gain greater stakes of control for its people in the face of oppression, have many tools at their disposal when employed properly renders a serious level of capability. That capability drives the fight. It creates those targets of opportunity that the guerrilla must exploit. A linear fight is a zero-sum game and one which the guerrilla cannot afford to engage – the primary advantage of the security forces of a nation state are that they can afford, and in turn replicate, their losses as well as divert assets to their most productive means in suppressing the guerrilla force through dominance of the electronic spectrum. Further, for the guerrilla, he must fully understand and embrace the role of signals intelligence to understand how he himself may be exploited. Know your weaknesses and you will recognize your strength.

In *The Guerrilla's Guide to the Baofeng Radio* I described the Guerrilla as existing inside the reactionary gap of a more sophisticated adversary. That reactionary gap is recognized as the lag between a professional force tasked with stability of an area to react to the threat posed by the guerrilla. Communications in all forms is a major, if not the most important, component to both creating and exploiting that reactionary gap. For the conventional security force, their own communications must both be streamlined for expediency but hardened against an unpredictable foe's attack. The conventional security force, an agent of an oppressive regime not of the people, works within the doctrine of the equipment he is provided. In many cases it presents a technological superiority in the outset, while simultaneously limiting creative thought.

Captured ISIS Yaesu VX-6R, a multipurpose tool for communications and signals collection.

They do, however, hunt for and based upon that metadata.

For the Guerrilla, his goals are first and foremost social change, with a number of advantages over the security forces seeking his defeat. He does not have a rigid doctrine from which to work. He is not married to conventional equipment nor is he encumbered with the limited working knowledge of the equipment that a conventional force provides its troops. There are no parameters for which he is limited in thought and practice aside from the overarching doctrine to guide him in his objectives. His equipment is most often sourced commercially which, as the conflicts of 2010 to the present have shown, enables innovation at an ever more rapid scale. That innovation has presented a critical challenge to conventional forces and has, at least as of this writing as it pertains to counter-insurgencies, won every time. In this new era of common, off the shelf warfare, conventional equipment has been found lagging behind even further than usual with the consequences at the feet of the troops tasked with facing down an army of the people. Lest the Guerrilla never forget this nor take it for granted, he will always find himself at home.

Understanding the Reactionary Gap

YPG Fighter in Rojava using a Yaesu VX-6R.

Understanding that the Guerrilla himself is not a disheveled, often impoverished fighter by choice but rather by necessity, his best attributes are his wits followed closely by the intelligence value provided to him from a broad variety of sources; everything from local rumors generated by informants to skimming social media to communications analysis in real time. These are used to direct activities of the Underground, whether that is in a direct action or supporting task tole. The Guerrilla force must operate where the enemy's capability is low or lacking. As it pertains to communications, this is only accomplished through putting a premium on understanding an enemy's collection capabilities both at the Tactical and Strategic levels. We do this through signals intelligence.

This creates the **reactionary gap,** or, the lag in time between actions a Guerrilla force takes and the reaction of the security forces. Of those intelligence sources, signals intelligence is perhaps the most valuable, followed closely by human intelligence, and in the modern era signals intelligence tools and capability are rapidly exceeding the tools and, more significant, the knowledge of proper implementation, fielded by nation states. The purpose of this book is exploring the role of intelligence, the tools required, and the techniques to properly employ them.

The quote by General Hayden, then Director of the NSA at the time of his statement, is a critical one for both the prospective guerrilla and the security forces assigned to hunting him. That metadata, containing who is sending the transmission, what it contains, how it is transmitted, and where it is received, is how every intelligence apparatus has targeted persons of interest since the advent of battlefield communications. While most often that context pertains, nearly exclusively, to cell phones, the same metadata applies to transmissions of any type; analog or digital radio, phones, email, social media history, or even person to person contacts when proper personal security is not observed. That data is a weapon in an of itself, identifying the unique signatures or patterns of a specific group. The trick, then, is figuring out how to find it.

The Baseline, Patterns of Life and the Social Network

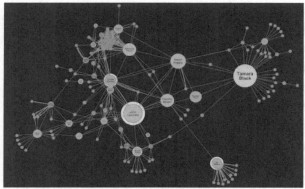

Social Network Map. *This applies to all forms of communications.*

Intelligence work, at its heart, is a puzzle of comparing knowns and unknowns with the end goal being exploitation of the target. Those knowns, whatever they are, create what is referred to as a **Baseline**. It is through that lens we define what is normal activity for a place. Once we understand what constitutes normal or routine, we build the first step in intelligence collection. That baseline allows us to eliminate what very well may be erroneous data sets – in other words, what to ignore – while turning our attention to more productive targets. Whatever violates the baseline becomes significant, or having intelligence value. There is some reason for the change in that baseline. Much of signals intelligence relies on the recognition of that baseline for a variety of reasons, depending on the intended role. One could be simple avoidance, as to not draw attention to your own communications, yet another violating that baseline to create a disruption. It is very much case-by-case, but creating the baseline is the foundation of any intelligence gathering task.

In recognizing the baseline we turn our attention to **Patterns of Life**. What we do, when we do it, and the standard activities of the day constitute the patterns of life of a target. Much of our intelligence work in Afghanistan targeting the arms smuggling network of southern Afghanistan, especially early on, hinged upon patterns of life surveillance missions. In these we created a timeline of events for a given target – when they woke up, when they ate, when they conducted their daily activities (*and what those were*), and when they concluded the day. This served a dual purpose – to know who came and went and when they did so, and when was the best time to exploit the target. This is no different than any other predatory behavior. The mountain lion stalks its prey, sometimes for days, learning every details of its movements from afar before striking.

Patterns of Life are also electronic. In the context of signals intelligence collection, radio signals (*and when they're emitted*) in certain frequency ranges can very well be action indicators. A common criminal activity is to have a police scanner monitoring the local PD radio frequencies as part of daily life in a trap house. Skilled criminals know the codes dispatchers and officers use, and listen for activities that might interrupt their activities. A guerrilla is no different when concerned with the local security establishment, constituting its own pattern of life. In another context a small team patrol may be using a specific type of encrypted radio for inter-team communications. While the tools to break that encryption may or may not be available, the presence of those signals violate both the baseline and the standard patterns of life. This is an **Action Indicator** – while we may not have the ability to break the encryption, the signals itself (*and its point of origin*) is far more important. Unmitigated, this can get a small team not practicing radio discipline during movement to the target killed.

The creation of a competent picture of the patterns of life of a target requires a serious level of patience and dedication to mission. For both the guerrilla and those tasked with hunting him, it can be frustratingly slow, mostly boring, inter-sped with brief moments of excitement breaking the mundane. But that dedication creates those victories, no matter how small, in the layer cake that is unconventional warfare. Understanding how patterns of life is collected and exploited tells us much in the way of mitigating the information collected on us at a given point while simultaneously hardening us as a target. In the guerrilla's context, it preserves the objective.

Patterns of Life gives way to the creation of a **Social Network** of the target. Who are they talking to and when, how frequently, and the nature of the association are all critical questions. Not unlike a detective tracking an organized crime outfit, the social network of a target creates a web of close associates and potential follow-on targets for exploitation. In the signals intelligence context, every aspect of the communications between the target and the recipient must be taken into consideration. Most frequently this has been done through the use of cell phone data, but is also easily created utilizing radio signals. One of the best examples was the successful exploitation of ISIS communications utilizing Digital Mobile Radio (DMR), each with their own unique identifiers, creating a metadata map which was then in turn used to direct both airstrikes and ground raids on key targets from the point of origin (POO) of the signals themselves. This is one example; callsigns, frequencies, and voice analysis can also provide this data, albeit over a much longer period of time. For that reason, over the duration of the Cold War, clandestine stations transmitting number code groups required no response to maintain the obscurity of the social network of its intended recipients.

Each of these must be taken into consideration in all aspects of intelligence collection, whether one is collecting information for future exploitation or seeking to mask their own organization. Who are you talking to, why you are talking to them, and the frequency of the contacts all play into paining an intelligence picture. It may sound complicated, and it is. This is however the way intelligence itself, regardless of type, is crafted. Failure to recognize its importance in the many roles creates those pitfalls that could end a social movement just as its gaining traction.

Intelligence Domains

Intelligence as a craft is broken down into a five task-specific domains. Human, Counter-Intelligence, Imagery, Measurement and Signature, and finally, Signals. These are broken down according to the diverse skill requirements of each and should be considered categorically both when information is being collected and when requests for information are being generated. While there is an overlap between each, they maintain several degrees of separation.

- **Human Intelligence (HUMINT):** The collection

and refinement of information utilizing human assets; ie, informants, saboteurs and spies.

- **Counter-Intelligence (CI):** Activities revolving around hardening oneself against intelligence threats; ie, compartmentalization and sanitation of sensitive information.
- **Imagery (IMINT):** Information and analysis provided by photographs of targets of interest, maps, or videos.
- **Measurement and Signature (MASINT):** Information collected based on specific patterns of behavior or the use of specific equipment.
- **Signals Intelligence (SIGINT):** Collection, analysis and exploitation of all means of communication both electronic and physical. Includes devices, operating modes, and enemy capabilities.

Intelligence as a topic, both for a conventional military or paramilitary force and for the Guerrilla, must attempt to satisfy each of these domains. It is a gross underestimation to divert the entire workload to one or two dedicated assets. Intelligence must become the primary goal. A special emphasis must be placed on SIGINT, especially in the field, due to its nearly seamless integration with ground forces when targeting an enemy in real time. While each are co-equal in importance, both HUMINT and SIGINT play a major hand in directing operations as the develop. As the Global War On Terror (GWOT) would force us to recognize, an adept Commander places a premium on streamlining SIGINT assets to his combatant forces on the ground.

"The Goal of Intelligence is Exploitation"

Intelligence of any type must have purpose. The definition of intelligence is understood as processed, verified information pertaining to an enemy used for some sort of exploitative role. This can be either how an enemy is targeted or defining one's own operating practices. Without the former, we have nothing more than information that may or may not contain veracity; that is, factual basis coupled with purpose. Without the latter, we've wasted our time. Intelligence must have teeth; that is to say, we must be able to weaponize that bit of data to our own advantage. The very nature of intelligence then is nefarious. It is a weapon as much as a rifle, if not more so. Intelligence is collected then weaponized in the Intelligence Targeting Cycle, better known by its acronym, **F3EAD**.

Intelligence Targeting Cycle (F3EAD)

The Intelligence Targeting Cycle is the logical process by which we locate and fix upon a target. Much like COL Boyd's well known acronym OODA, or Observe, Orient, Decide, and Act for recognizing and engaging threats, the Intelligence Targeting Cycle is better known by its acronym **F3EAD; Find, Fix, Finish, Exploit, Assess,** and **Disseminate.**

- **FIND:** A target is defined then located.
- **FIX:** Surveillance is emplaced in order to define Patterns of Life
- **FINISH:** A method of attack is initiated on the target
- **EXPLOIT:** Collection of data for future use / weaponization
- **ASSESS:** What was the outcome and / or effect?
- **DISSEMINATE:** Refinement of more successful methods of attack

F3EAD appears simple on its face, as all logical methods should. Simplicity is deceptive. It must be pointed out that F3EAD should be considered at all times both by the intelligence collectors in the field and those tasked with its analysis. If, for whatever reason, the information being collected does not satisfy one of the steps of F3EAD, the persons involved are wasting their time and need to reconsider their actions.

Intelligence is, above all others, the most critical task of an unconventional or guerrilla warfare underground. The Guerrilla exists in the seams and gaps of an oppressor force, striking those targets of opportunity before the oppressor force can react. The technique, not necessarily the attack itself, is what is most important to recognize. Further, the role of intelligence provides those very seams and gaps from which to exist much as the analogy of the *War of the Flea*, defining everything from the manner of attack to understanding the effect to defining one's own operational parameters. How would a fledgling guerrilla force know, for example, which frequencies to operate in creating a Signals Operating Instructions (SOI)? How would they know which modes and methods are best suited to their operational environment and which are at the highest risk of compromise? Only through careful consideration to intelligence can any of these parameters be ascertained.

F3EAD as a logical process provides those answers both defensively and offensively. Signals Intelligence (SIGINT) becomes as critical to mission planning as it is in understanding how to attack a foe. The guerrilla force recognizes that they are in a position of inherent disadvantage in terms of dedicated personnel, especially in the early phases of a conflict. A conventional or occupation force can afford to absorb loses in men, weapons and equipment. Equally true is the dedication by which they can divert resources to a broad number of roles. The guerrilla cannot afford to do this, but, on the other hand, recruits specialists circumventing a rather long training pipeline. However, as mentioned, the conventional forces work within the parameters of their equipment and evolve slowly to the realities of a contemporary battlefield. The guerrilla, as a survivor, adapts quick. F3EAD, as a process, is a major part of that rapid evolution.

Collection and Analysis

There are two distinct categories of intelligence personnel; **Collectors** and **Analysts**. *Intelligence Collectors* are those personnel operating in the field whose job entails observation of a target with a specific requirement. That requirement, often known as a **Request For information** (RFI) is generated by the Analyst. *Intelligence Analysts* are tasked with mapping and assessing the patterns of behavior, or those aforementioned **Patterns of Life**, created by a target. Analysts sort the information collected in order to brief the Commander and generate new RFIs for the Collectors in the field. Ideally, the flow moves quickly and relatively seamlessly. A good collection and analysis team generally have a flow, and as time and familiarity progresses, a collector gets an eye for what's relevant and what's not.

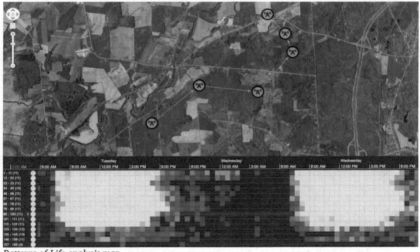

Patterns of Life analysis map.

Although there is a certain level of overlap between the two roles, and certainly there is value in that strong working relationship, it must be pointed out that the two roles are very much mutually exclusive. A Collector's job is to collect, not infer nor analyze. He must send up raw data, nothing more and certainly nothing less. I've known personal cases where intelligence Collectors began to analyze the information collected on the ground before sending up the situation up to a higher authority. Not only did the Commander lack a degree of situational awareness, much was lost in the transition. Fast, factual and faithful are the watchwords of a competent intelligence collection team.

An Example of Collection in Real Time

Captured equipment from a Taliban High Value Target. Note the Radio. My team captured this equipment after eliminating the target.

In Afghanistan our primary role was intelligence collection through physical, eyes-on-the-ground means with the capability for exploitation in real time. In short, we were the physical eyes on the ground collecting what was tasked of us as is the traditional role of *Special Reconnaissance* (SR). While we had numerous capabilities through our own training pipeline, we were assigned *Multi-Function Teams* (MFTs) that had specifically trained intelligence collectors assigned to us, specifically in human (HUMINT) and signals intelligence (SIGINT). We were the eyes and the teeth, and paired with HUMINT collectors, whatever was collected after interdicting a target could be processed in a more streamlined way to get back to the analysts. It made life much easier and our productivity rate that much higher, but our real success came from being paired with the SIGINT collectors.

Our SIGINT collection team worked on two primary targets: two way radio and cell phone traffic. Understanding how the Taliban communicated, by two way radio at the tactical level first and by cell phone at the strategic level, data was collected and exploited from their chatter and used in a number of ways. The first method was through Direction Finding (DF) of the Point Of Origin (POO). This was used to geolocate the target. The second method was through analysis of the intercepted traffic. Through both DFing and traffic analysis, our SIGINT team enabled us to strike more competently and effectively. At least for a while the Taliban had a hard time catching up – eventually they did, however, as insurgents always do – but the lesson in streamlined intelligence collection no doubt reigns true. The faster, the better, the more effective, and particularly so for those operating in the traditional Unconventional Warfare (UW) paradigm.

We had a mission in the mid-winter to clear a well-known mountain pass linking the Arghandab River Valley to another valley leading to the border with Pakistan. American forces had rarely traversed this pass during favorable times, much less in the dead of winter. The Taliban used it as a conduit for arms and explosives, paying little attention to any threat from the ground but rather from the sky, and in particular, drones and helicopters running sorties. They were content to reside in their caves, warmed by fires, relaxing over the winter for the coming fighting season in the spring. They were not expecting an attack from the ground and were ill-prepared for it, on top of the shock that Americans were moving in their valley without their vehicles. This generated an unusually high amount of radio traffic that was recorded and translated on the fly – they were cold and scared, unable to light their fires lest they be spotted, and were in no mood to fight.

The mission was a high success. Several targets were interdicted without much of a fight and we took zero casualties. A big part of that success was owed to the solid work our SIGINT team did on the ground with us, rather than having to rely on more remote assets. Knowing they were at a disadvantage made us desire the fight that much stronger, despite the near-zero temperatures enduring over multiple days. The very same success in the modern era of Guerrilla Warfare can be had with relatively inexpensive, commonly available equipment. This book will describe in detail that equipment, its operation, and proper employment.

Processing Potential SIGINT Targets

Everything that is collected in any of the intelligence domains, tells us a story. To a certain degree, humans all follow the same patterns. We have needs, we have hierarchy. The very purpose of targeting communications has a goal to mitigate or reduce that target's threat to us. In pointing that out for the Signals Intelligence context, we must seek to answer questions in this order:

- **What is my target transmitting?**
- **Why is my target transmitting?**
- **What is the purpose of his communications?**
- **What is his place in the hierarchy of his organization?**

- **What is his level of training and / or competency?**
- **What is their threat to my operations?**
- **How might I exploit this target?**
- **What, possibly, is the second and third order effect of exploiting this target?**

These questions provide that metadata, as General Hayden would point out, that enables further targeting for exploitation. Each of these questions paint for us a picture of our intended target. How we hunt him, how we can exploit him, simply based on the way in which he communicates. This begins with the source of the transmission and that metadata becomes the physical **Point Of Origin** (POO) of the signal, any observed identifying callsigns, and the properties of the transmission itself (*frequency, duration, digtial vs. analog, etc*).

One of the most important factors, at least in my experience, is not simply resigning ourselves to analyzing the traffic itself or even the geolocation of the source, although these are critically important, but determining the level of training and competency. A target with poor communications discipline, even if well equipped, tells us he does not understand his equipment and likely is inexperienced. These types of targets, for a guerrilla, are easy prey. Inducing any type of stressors or stimuli, whether that be an electronic or physical attack, may very well disrupt or disable his capabilities. At a minimum it creates that reactionary gap in which the guerrilla finds himself a home. But before those actions can be taken, collectively known as exploitation, the SIGINT collector must attempt to map every aspect of his target's capabilities. That begins with the SOI.

Signals Operating Instructions (SOI)

For the SIGINT collector, a valuable starting point is drawing out a blank Signals Operating Instructions (SOI) sheet. Even when partially filled out, this creates a roadmap for what we have collected and can answer many of the questions above. An SOI is a communications plan for a specific organization for the duration of an operation. For tactical communications purposes, a competent organization creates the SOI during the planning phase of a mission and is destroyed at the mission's conclusion – never to be used again – as a basic communications security (COMSEC) measure. That said, recognizing that human nature is inherently lazy, failure to understand even this basic practice can be a major windfall for the SIGINT collector.

In Afghanistan the Taliban married themselves to VHF repeaters for tactical communications with little consideration to COMSEC. It was relatively simple to exploit the signals, but what was difficult was attempting to map the hierarchy of leadership. One solution was to simply destroy the repeaters themselves. No repeaters, no communications. While that seemed logical, we quickly realized that their communications network simply adapted to the situation by communicating peer-to-peer, and we had simply blinded ourselves and made our jobs that much more difficult. We had to now hunt for the targets we previously had identified through paring the data previously collected to the active cell phone traffic in the area. The point is, we would have been better off simply not making them aware of our presence until it was time to interdict the target. Had they not married themselves to a repeater network to start, we may never have truly mapped their communications capabilities nor their leadership in real time. Their SOI, to whatever degree they implemented it, would have been in a constant state of change and more difficult to exploit.

Sample SOI:

Signals Operating Instructions (SOI)

Primary [FREQ / MODE]:
    PROWORD:
    TX:
    RX:
Alt. Freq [FREQ / MODE]:
    PROWORD:
    TX:
    RX:
Contingency Freq:
Emergency Signal:

Callsigns:
    TOC / Control:
    Element 1:
    Element 2:
    Support/Recovery:

Challenge / Password:
Running Password [DURESS]:
Number Combination:
Search and Rescue Numerical Encryption Grid [SARNEG]:

    K I N G F A T H E R
    0 1 2 3 4 5 6 7 8 9

    0 1 2 3 4 5 6 7 8 9

*** SARNEG first number rotates with last numeral of previous date***

COMMO WINDOW SCHEDULE:

	1	2	3	4	5	6
7	8	9	10			

- **DAY:**
- **NIGHT:**

Using the SOI to Process SIGINT Collection

  In the SOI above, the frequency list is created using a PACE plan: Primary, Alternate, Contingency, Emergency. From there, those specific transmitting modes are indicated, along with the codeword used to identify which frequency set the target is operating on. In the event they are not using one, this can be left blank. The transmitting frequency is indicated by TX, the frequency they're receiving traffic is listed as RX. This is called duplex mode and is a simple COMSEC procedure that creates a simple way to frequency hop. This also prevents continuous traffic on one frequency. From there we map any potential callsigns that have been heard along with their purpose or indication of hierarchy, if known. Finally, we map the authentication code they may be using in the form of a SARNEG and logging the times of the observed transmissions with the communications window list.

The SOI provides us with a logical roadmap to list whatever information we collect – the frequencies, the callsigns of the hierarchy of organization, the observed modes (*if any*), and even a timeline for listing the target's transmissions. They may or may not have the level of sophistication to use the SOI in its complete form. If not, use what is relevant. What this enables us to do is paint a logical picture of what is being transmitted in an effort answer the questions listed above, creating knows from unknowns. While there may be other means of logging the data, utilizing the SOI allows for a logical process and a standardized way to hand that information off for further analysis, preventing any relevant information from potentially being left out.

Weaponizing SIGINT – Scraping Metadata

The process of intelligence cultivation, as pointed out, must be centered on the eventual exploitation – absent that exploitation, we are simply wasting our time. SIGINT can provide, above all other domains, real time targeting information seamlessly integrated into the small unit. This manual will address the equipment required, the process of collection, and the various means of exploitation in depth. It is important to realize that intelligence is very much a dagger that cuts two ways; what can be used against an adversary is absolutely being used against you. Communications is the least understood of all enablers on the battlefield, with SIGINT, the other side of the communications coin, often taking a backseat in the minds of many. It is however the fastest way to both target and be targeted based on the RF emissions and understanding this truth enables the guerrilla to exploit a technologically superior foe with relatively cheap equipment. Training based around this reality must become the goal. To those who seek simple solutions requiring no work, the book will illustrate the rewards reaped by the lazy. To those seeking to understand and build their own capabilities, however, you will come to understand not just how to conduct the art of signals intelligence collection but gain greater understanding in how to operate more securely in the era of ever-increasing electronic surveillance capabilities.

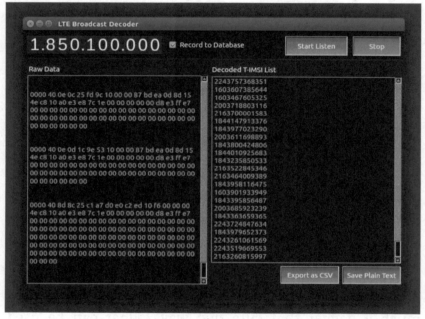

Metadata goes far beyond what most people recognize – we most often think of metadata in terms of cell phone IMSI, IMEI and SMS data, email headers, footers and addressees, while failing to realize its literally any piece of data that can be used to positively identify or otherwise exploit a target. This includes callsigns, modes, radio equipment identifiers (in DMR and P25 especially), the frequencies you and your team are operating on and yes, your voice itself. All of this creates a composite package that can be used against a team on the ground at the hands of a capable Signals Intelligence team. It is the purpose of this book to both inform the reader on those methods, recognizing that they can both be of benefit and can also avenues of exploitation.

SIGINT – The Defining Weapon

SIGINT is quickly becoming the defining tool on the battlefield by which every other weapon is becoming dependent upon for targeting purposes. As the electronic domain becomes ever more cluttered with better mousetraps, so too comes the ways in order to exploit them. In all aspects of Signals Intelligence it becomes imperative to understand these concepts not simply for your own exploitation capability but in building a better awareness of what will be fielded against you. And for that reason this book exists, both as a guide to creating your own capabilities, but recognizing the threats. In saying that much of the same capabilities, or at least comparable in role, is commonly available. As General Hayden succinctly pointed out, we do indeed kill a lot of people based on metadata. Be the Hunter, not the Prey.

CHAPTER 2. TOOLS OF THE SIGINT TRADE

The craft of intelligence collection, like any other trade, requires several pieces of equipment to paint a comprehensive picture of the electronic spectrum of a given area. The inexperienced or amateur collector will default to assuming everything monitored needs to be decoded or deciphered in order to be exploited. This is not so. In most cases simple awareness of a transmission being made is enough to exploit at the tactical level.

Like with anything in the communications world, many are quickly overwhelmed by the options that exist and the technology itself. A competent signals intelligence package can be as simple as a map and compass, a communications receiver, a Baofeng radio, and possibly a Software Defined Receiver (SDR) in the hands of a trained user. Those three tools can create a basic level of capability for very little money. On the other hand, expensive and elaborate setups may offer more in terms of overall capability, but absent a skilled operator, will be of limited use. The answer is found in the skill versus the equipment itself.

Requirements of the SIGINT Role

Signals Intelligence as a task has two primary requirements:

1. **Situational Awareness**
2. **Tactical Signals Exploitation**

Situational Awareness (SA) is, as the name implies, maintaining awareness of all things in a working environment. This could be everything from knowing the local weather report to knowing when the shift change is coming up for the dispatcher's office and everything in between. Most often law enforcement and emergency service traffic falls into the category of SA – it is not traffic that is usually actionable, but good to know nonetheless. Assets dedicated to SA are usually in a fixed site (*sometimes known as a Listening Post, or LP*) such as a safe house of guerrilla camp. The equipment used consists of a scanner, either a handheld or base station unit, and a dedicated receiving antenna that is either permanent or semi-permanently installed.

Tactical Signals Exploitation (*TSE*) on the other hand is the weaponization of the data found. This is the role of the teams going on patrol and directing the fight on the ground. This includes everything from radio direction finding (*RDF*) the source of a signal to establishing that target's pattern of life to cryptanalysis to jamming them, and everything in between. Exploitation, in this context, is the conducting of warfare, physical, electronic and psychological, on your intended target. Our equipment, whatever it is, has to be able to fill these roles for us. Knowing how to press specific types of gear into service presents a large advantage to the guerrilla, who often does not have the luxury of large budgets and government assets. Much of these tasks can be done with common, off the shelf equipment with surprisingly little money invested.

SIGINT Soldier configuring Wolfhound RDF device for targeting VHF & UHF signals.

TSE teams are a critical asset for the small unit in the field. As discussed in Chapter 1, in Afghanistan we had a SIGINT team permanently attached to us tasked with conducting electronic surveillance. They were fully integrated into the team and created a high degree of success. The same can be said for the role of Special Operations Team – Alpha (*SOT-A*), SIGINT teams integrated into the Special Forces Groups, the Intelligence Support Activity (*ISA*) in its many incarnations supporting Tier 1 Special Operations and the USMC's model of pairing SIGINT Marines with Scout Snipers for streamlining specific targeting information. All of it owes to the efficiency of integrating SIGINT at the small unit level.

SIGINT soldiers getting a line of bearing on a signal with a Yagi antenna.

For a guerrilla force the same is absolutely true and is even more important, at least in my opinion. The role is further enhanced both by the multitude of common, off the shelf technology and the freedom of outside the box thinking. The guerrilla's best assets are speed and surprise, with both relying upon efficient targeting on while on a patrol or for the security of their safe houses and guerrilla bases. While many may view signals intelligence as an afterthought, it is in fact one of the most critical tasks. One of the most streamlined examples I can think of is simply pairing a Sniper with a Communications Receiver and keep a log of targeted frequencies. When the target's signals are detected, and a **Line Of Bearing** (*LOB*) and/or geolocation is established, the Sniper now knows where to get in position to take his shot. Exploitation in real time.

Let's expound on this scenario. When I teach small unit tactics to civilians often the question lingers in their minds "how do we know where to place an ambush?" and that's a fair question. Its not one that can be answered aside from the one provided in the book, "along natural lines of drift that the enemy is taking". But it is an answer that SIGINT can provide. If you know anything at all about the signals your adversary is emitting, say, in this example the **Russian Azart radio system (*27-500mHz, TETRA digital protocol*)** and you detect it and direction find it in real time, you now know where the enemy is operating and from there examine your map for natural choke points to place that ambush.

Sourcing SIGINT Data

The most important question many people have before sourcing any equipment is where to locate the data itself. If you're not familiar with communications in general this gets confusing in a hurry. Fortunately there's several resources that can help break this down and one that you'll likely be constantly referencing. The first one on the list is **Radio Reference** [radioreference.com] which breaks the data down to the individual county level. This not only gives us a list of data for an area but creates that baseline of what's normal. This includes everything that breaks squelch in a place – everything from emergency services to water meters to amateur radio traffic to cell phone towers. So, for example, if I'm intercepting an unknown digital signal on 155.355 and I look up my county, it turns out to be the local Sheriff's Department dispatch. Might be of interest, might not. But we could file that away as certainly good to know.

Radio Reference is a great tool but, like all things, has a couple of drawbacks. We don't always know if its up to date – that information has to be verified – and it also requires a subscription to unlock the full features of the site. That's all good and well for the situational awareness tasking, but its not going to give us much else than the published data. For targeting or exploitation purposes this really isn't doing us too much other than helping to create the baseline for an area.

The next online reference tool that I find myself using most often is SigIDwiki [sigidwiki.com], which is an online database of all known signals and digital modes. It is 100% free and fully downloadable with the with their software suite known as Artemis 3. This is, at least in my opinion, the most important tool in the SIGINT arsenal because it not only breaks down the known data of specific modes, but what frequency ranges you'll find them in and who's using them. Perhaps most important is the waterfall graphic that's included for spectrum analysis which we'll be talking about in the next section.

While there's other sources of data out there, these two are the ones I keep coming back to not just for their ease of use but the reliability of the data itself. If I'm headed to a new place the first task is creating that baseline to understand the knows and in turn find the unknowns. That's difficult enough, and those two references make life just a bit easier.

SIGINT Collection Equipment

Now that the purpose and integration behind intelligence collection are understood, we have a number of tools at our disposal, each having their own place in the arsenal. While there is some overlap between each, first its important to recognize the role behind the tool itself. A competent signals collection package contains three specific categories of equipment:

1. **Physical Monitoring**
2. **Spectrum Analysis**
3. **Signals Exploitation**

Physical Monitoring, as the name implies, concern the devices that emit audio output to the operator whatever they receive over the air. This can include captured enemy transceivers themselves, scanners and communications receivers. All are separate tools with a specific role. Everything that emits RF has a specific signature, regardless of whatever means are used to encode and/or obscure the signal's contents. Whether the physical monitoring device can decode / decrypt the transmission or not is, often enough, irrelevant at the tactical level. Time wasted trying to decrypt an intercepted signal a mile away can be better utilized preparing an ambush along their direction of travel or an early warning to their presence.

Static TV used as an early warning to detect SINCGARS traffic.

A perfect example was from my first tour in Iraq. The al Qaeda Iraq (AQI) insurgents operating in the Rashad Valley region utilized TV sets as a type of early warning device. They noticed that our ground communications, which operates in the 30-88mHz range, interfered with their TV reception. Broadcast TV is also in that frequency range. When we got close and transmitted, the TVs themselves would either get snowy or the static pattern itself would change. The SINGCARS radio system operates in the 30-88mHz range, frequency hopping in set intervals within that spread of frequencies. Television receives between 54-88mHz, meaning that the SINCGARS may interfere with TV signals. Just so happened they did.

What they began to do was cut on a TV and leave it on a blank channel with the volume muted. When the static changed, they knew it was time to move. There was one raid in particular on a small village that had been, allegedly at least, been used as a safe house. By the time we were on target the place was empty, save for the TV left on in the corner of one of the mud huts.

While basic by anyone's standards, it worked and owed to our lack of communications discipline while on the move. We may have had far more sophisticated equipment, but they thought several moves ahead. It owes to the saying that warfare follows a cycle – low tech defeating high tech, high tech creating counters for low tech solutions. In this case it was as simple as a muted, static TV. But it is, however, the perfect example of physical monitoring.

Physical monitoring has three categories of equipment for SIGINT collection. Scanners, both analog and digital, communications receivers, and transceivers themselves. Each are different in specific role and application, but do have some overlap, especially scanners and communications receivers. But that said, they are each of equal importance in the overall package – and while I have my preferred loadout, each tool may accomplish the task.

Spectrum Analysis is equipment that gives us a physical display of a wide spread of the radio spectrum.

Spectrum analyzers, SDR and frequency counters each fall into this category. They quickly give the user a visual indicator of anything that is emitting a signal within the device's receiving range. Spectrum analysis does not normally give us audio output, but can, in the case of certain devices, give us a visual indicator of the radio waves being emitted. This can immediately tell us what type of transmission is being intercepted.

Spectrum analyzers are incredibly useful due to their small size and ability to monitor wide portions of the spectrum at once. As the technology has advanced over the years, these have found themselves on an ever more important role both on the ground and in the air. Aerial platforms, such as Intelligence, Surveillance and Reconnaissance (ISR) drones and manned aircraft utilize spectrum analyzers to identify targets of interest prior to any physical monitoring. This makes both the mapping of the RF landscape, ie all of the signals being emitted over a given area, and identifying targets of interest quick. That said, the expense of the devices themselves are coming down substantially, enabling access to a level of sophistication to the guerrilla on a dirt cheap budget.

Signals Exploitation are the devices used to conduct electronic warfare on a target. This is done in a wide variety of ways, from jamming techniques in an attempt to disrupt enemy transmissions to conducting psychological warfare to degrade his morale. This can include transmitters themselves, whether off the shelf or captured, and recording devices to capture the signals intercepted.

Scanners

The first stop for most when it comes to signals intelligence is the common scanner, usually referred to as a police scanner. The traditional way a scanner works is to scan a specific range of frequencies for specific services – these are portions of the frequency spectrum where types of communications are assigned. Scanners come in two types; **analog** and **digital**. Both types generally scan spreads of a frequency spectrum extremely fast. But with that said, the equipment is built around the role of scanning certain frequency ranges alone, and not as a broad spectrum receiver.

Analog scanners only receive whatever signals come over a given frequency. That is to say, they do not have an ability to decode anything that is digitally encoded. They do however still receive whatever traffic is coming over that given frequency, and that can be valuable to know in and of itself. Many consider analog scanners to be obsolete, especially if the goal is monitoring local law enforcement, many of which have went to digital systems for efficiency. While that opinion is not wrong, it does however create an opportunity to buy analog scanners at a bargain price.

One of the most common scanners is the Uniden Bearcat series which comes with a feature known as *Close Call*. Close Call is better known as close proximity frequency capture, meaning when in this mode the scanner will tune to the closest strongest frequency in use. For example, if a target is using an FRS radio, like the Motorola Talkbouts, and is in close enough proximity, the scanner will automatically tune itself to the frequency they're operating on. It is one of the more impressive features of the scanner, whether analog or digital, and is well suited to integrating on a patrol.

Digital Scanners have the capability of receiving

both analog and digital radio traffic of a given area, provided it has the correct protocol for the digital mode. The most common digital mode in public service in the US is P25. P25, short for APCO Project 25, is currently in its second phase. It is a digital encoding method introduced by Motorola in the mid-1990s to first increase efficiency of emergency communications in large municipalities and second to offer an increased level of security. P25 became ubiquitous in many jurisdictions in the US and is still in widespread use today.

Scanners designed to receive digital traffic usually prompt the user to enter the zip code of their location when the scanner starts up. This pulls the data from a built-in memory bank which can usually be updated. This pulls all of the communications data, including frequencies and the names of the agencies using it, without much on part of the end user. It is simple to use but has some drawbacks, depending on the end goal.

For basic the situational awareness task this is fine. You'll be able to monitor the local emergency service traffic. You will not, however, know whether that traffic is being transmitted in real time (many jurisdictions operate on a delay) nor will you know the specific frequencies in use, of which there will be several – you'll only know the frequency of the primary dispatch transmitter. Further, a Digital Scanner will only be capable of receiving the traffic it has the capability of decoding, and this is based on the zip code that was input when started up. While most have the ability to scan specific frequency ranges in addition to the pre-set frequencies, there may be much that is missed.

Newer models, such as the Uniden SDS-100, are seeking to close the gap in application by adding in more capabilities, but this comes with a significant increase in expense that can be filled with other equipment at less cost, if it is even required at all. *Keep in mind that at the tactical level decoding / decrypting digital transmissions are not usually necessary for exploitation, especially if it is known that that specific method is used by a group already recognized as hostile.* Their presence and patterns of life are indicated by the specific mode they're using which in turn alerts you to their presence. At the tactical level, knowing that they cannot be too far away, a better option for exploitation becomes simply closing with that signal and killing whoever is emitting it. Or avoiding it, depending on what your end goal actually is. But either way a competent adversary is rotating encryption methods at a regular interval and dedicating the time and effort to breaking it is most likely a fruitless endeavor. We can identify the mode our target is working in through the waterfall image on an SDR display or the Spectrum Analyzer and then gain an audio clip (and Line of Bearing to the signal) with a Communications Receiver. This limits the utility of a scanner for much beyond the Close Call role or Situational Awareness mentioned above for capturing whatever signals are being emitted in close proximity.

Communications Receivers

Communications receivers are often confused with scanners. Both scan frequency ranges and are capable of monitoring and mapping received traffic. But this is where the roles begin to diverge. Communications receivers, sometimes referenced as wideband communications receivers, are designed to receive everything in the radio spectrum from the AM broadcast range (*500-1800kHz*) to 1000mHz or sometimes higher, without the gaps in coverage that scanners have. There is a reason for this. Scanners are designed for optimization on specific frequency ranges where emergency service, aircraft, and utility traffic are commonly transmitted. Communications receivers are built to receive everything. That said, while scanners are a tool for situational awareness, a communications receiver is better suited to finding and mapping the frequencies in use themselves, along with giving us an audio output of the signal.

This makes the communications receiver one of the best all-around tools in the SIGINT tool kit. In addition to receiving the signal, it gives us the frequency, an audio output (*regardless of a digital or analog transmission*) and a signal strength indicator which is critical to radio direction finding. They are more utilitarian in purpose and application, in addition to generally being faster to employ when paired with a Spectrum Analyzer or Frequency Counter.

Two of the best communications receivers are the AOR AR-8200 and the Alinco DJ-X11T. The former was in common use in Afghanistan as a signals collection tool paired with a frequency counter to quickly capture whatever frequencies it detected. Both communications receivers receive from 50kHz to 1300mHz. The dial pad on the front panel allows for quick entry of a frequency. The Alinco DJ-X11T is similar in operation, with a built in feature known as F-Tune that works the same way Close Call does in Uniden scanners. This makes the Alinco an incredibly versatile tool for use at the Small Unit level.

One other major advantage to the Alinco is the ability to scan two separate frequency ranges simultaneously. This is critically important when attempting to map the communications of a more sophisticated adversary using multiple frequencies at once. For example, in *The Guerrilla's Guide to the Baofeng Radio* I cover configuring the radio to transmit on one frequency range and receive on another. This can make interception difficult, especially if using burst transmission techniques as described in that book. The ability to scan multiple frequency ranges at once is a major equipment advantage for mapping the operating frequencies of an adversary.

Quick Start with the Alinco DJ-X11T

The Alinco, like most Japanese radios, is a fairly complex tool to get the hang of at first glance. There's a full function numeric keypad with MAIN, SUB, SCOPE, SCAN and V/P/M buttons. Every button has three functions which are controlled by the large button on the left side of the radio that is where the push to talk (PTT) would be on a two-way radio. Below that button is the monitor button (opens the squelch to listen to weak signals) and at the bottom is the power button.

On the top of the receiver you'll see two knobs that control volume and frequency tuning. The one on the right controls the bottom frequency, the one on the left controls the top. The bottom half of each knob controls the volume of each of those frequency displays (they're independent of one another) and tapping the knob opens the squelch level option.

We want to keep the X11T in VFO mode. The VFO mode is a direct frequency entry – whatever is on that frequency we'll hear. These options are controlled by the V/P/M key. You can program the memory on these using Alinco's proprietary software, but I've never seen a need to do so. You're writing down the frequencies of interest and should have a quick reference at hand as it is. Keeping it in VFO mode allows us to dial in frequencies as necessary and scan entire portions of the band space. This is important when we're conducting our spectrum search.

The MAIN and SUB buttons control the top and bottom frequencies. When you hit MAIN, the top frequency gets larger on the display. You can now enter a new frequency or set it to scan by hitting SCAN. SUB controls the bottom frequency. When you press it the bottom frequency gets larger and you can enter a frequency or set it to scan. If you hold down the MAIN button the SUB frequency will go away, and if you hold down SUB the MAIN frequency will go away, allowing you to only listen to that one frequency.

To enter a frequency, punch the first number in megahertz (MHz), then the dot key, then the kilohertz

(kHz) after that. For example, if I wanted to listen to 151.820 MHz, I'd enter 1, 5, 1, dot, 8, 2, 0 then hit ENT to enter the frequency. If you don't do that the radio will default back to whatever frequency it was previously monitoring after a couple seconds. If you get turned around, just give the radio a couple seconds and then try again.

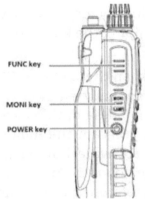

To enable F-Tune (Alinco's 'Close Call' function), first tap the menu button (FUNC) found on the side of the receiver then press SCAN. Pressing it once opens the F-COUNT feature, which is simply a frequency counter displaying whatever signals the receiver is picking up that second. I've found this feature to be pretty much useless, since it doesn't stop on any specific frequency and moves to fast to write them down. Press SCAN again to open the F-TUNE function. When the receiver locks on to a nearby strong signal it'll emit a loud beep and that frequency will now show up on the display. Now you're monitoring whatever frequency you found.

A word of caution about "Close Call", F-Tune and any other near field frequency capture feature: In my experience these work well for rapid tuning at the tactical level, where groups may be less than 1 km apart. For that purpose they're effective for at least mapping the frequencies that may be in the primary use. I have seen those in the "Intelligence" field advocate for their use alone as the epitome of signals intelligence capability. They're wrong. In my experience, this leads to a false confidence in your equipment especially if an adversary has any level of communications security (COMSEC) training. The proliferation of this feature is a convenient tool, but is easily defeated by both distance and receiver sensitivity. It is absolutely not a crutch or even a feature I've come to rely upon due to these pitfalls. The TinySA Ultra spectrum analyzer is far better for the rapid identification of frequencies in use and is a more reliable tool. The communications receiver's role then becomes entering that frequency data that the TinySA picks up.

AOR DV-10

AOR has a new generation of communications receiver on the market, the DV-10, which has replaced their old warhorse, the AR-8200. The AOR DV-10 is a handheld wideband receiver that takes the capabilities of the AOR-8200 and DJ-X11T and advances them into the world of digital communications. The radio can demodulate / decode every digital mode voice mode in common use today, including P25 & DMR. It was controversial when it first came out mainly due, at least in my opinion, to its cost. Early users had complained about frequency stability issues when using it to intercept DMR signals and this is not an issue I have been able to replicate. Since it was poorly received early on the units have to be purchased directly from Japan. But, in my opinion at least, its one of the best kept secrets in Signals Intelligence at the ground level.

The DV-10's display is similar to the X-11T in that it has the ability to monitor two frequencies at once. The DV-10 is also best run in the VFO mode, which is set by pressing MENU, then selecting VFO from the options, then selecting either A or B, then pressing ENT. You can alternate these two quickly on the display by pressing ENT. If you get turned around in the menus, press CLR to return to the home screen.

To set the listening mode, press MODE, then use the arrow keys to highlight which mode you want the frequency you're listening to to be set on. The default setting is automatic (AUT 1) but this can be hit or miss in my experience. You can pick between FM, AM, Upper Side Band (USB), Lower Side Band (LSB), CW (Morse code), D-Star, Yaesu System Fusion, DMR, P-25, dPMR, Alinco's proprietary digital mode, and trunked DMR (Tier III) reception.

To set the squelch, hit the SQL/MONI button on the side of the radio just below the power button. This highlights the LSQ (squelch level) on the display and you change this by rotating the bottom knob.

The direct frequency entry is exactly the same as the X11T – punch in the numbers for MHz, the dot key, then the kHz, then hit ENT to enter them. Its a simple set of controls to rapidly tune the receiver once a target frequency has been identified. You can use the arrow keys to shift the frequency up and down as well, while adjusting the tuning step (how many kHz it jumps with each shift) in the options page under MENU.

The last feature the receiver has is a built in recorder which is turned on by the large circle icon button in between the MODE and CLR buttons. You can access these recordings for playback by holding down the recorder key until the menu pops up. From there you can select the recordings to replay. This comes in handy during the surveillance and analysis of a target's communications.

The DV-10 adds a serious level of capability to the tactical SIGINT package in that it can decode most of the common digital modes out there. Although I'll reiterate that this is not always necessary and that it by no means makes analog communications receivers obsolete, its is however a great tool to have albeit at an expensive price. It does come in probably the most robust package of all of the tools covered in this book and is clearly marketed to government agencies. If you can afford it, the DV-10 is certainly worthwhile adding to the SIGINT arsenal.

Using Transceivers as SIGINT Tools

In addition to the tools already discussed, transceivers themselves can be used as SIGINT tools in a couple of ways. First, they make effective scanners for the given frequency range in which they receive. Second, they can be used as an exploitation tool to conduct electronic warfare.

Several handhelds on the market are capable of wideband reception, meaning they receive in the same wide range (*or close to it*) as a communications receiver. Kenwood's TF-6A and the Yaesu VX-6R are two very good, even if a bit dated, examples of radios with wideband reception from the AM broadcast band through 1000mhz. I do not consider them to be in the same category as a dedicated receiver, however. Transmitters in general do not have the same receive sensitivity, which may or may not matter at the tactical level, but it is still something to note. Further, Yaesu is fairly complicated to use and only scans a single frequency range at a time – making it somewhat slow. This is on top of the short battery life and somewhat poor signal selectivity. That said, it can and has been pressed into service as a collection tool.

The Baofeng, in all its incarnations, is another tool that should not be overlooked. It receives from 136-174 mHz (VHF) and 400-470 mHz (UHF), as well as 67-108mHz (receive only). As a scanner, it is hampered by a fairly slow scan speed. This can be sped up, however by pressing MENU then scrolling to MENU #1 (STEP). This option allows you to select the bandwidth of the frequencies while scanning. The default setting is 25kHz, which is a standard frequency spacing to keep two nearby signals from stepping on one another. Voice communications are 12.5kHz wide. Digital and narrowband voice signals are 6.25kHz wide. This means setting the radio to those settings will make it stop on the exact frequency, but leaving it at 25kHz will speed up the scanning capability without missing the wideband signals. Cutting it down to 12.5kHz narrows it down to catching the digital signals. It is still less than ideal.

The best way the Baofeng is pressed into service as a scanner is by loading the license-free frequency data into its memory. While that might be counter-intuitive to what's advocated on setting up a Baofeng for use in The Guerrilla's Guide to the Baofeng Radio, this is for scanning purposes rather than utilizing them to transmit. The FRS, GMRS, MURS and Marine frequencies are loaded into the memory of a specific radio set aside as a standalone scanner. This makes scanning those far faster on target and should serve as a warning not to use channelized data for your own operating frequencies. In the US, for whatever reason, many in preparedness circles advocate using channelized data for tactical and clandestine communications despite this fact. It is a major detriment due to both laziness and a severe misunderstanding of the duplicitous nature of communications for a prospective Guerrilla force. Be that as it may, channelized data in the VHF and UHF ranges should be a first focus for signals intelligence training of personnel. It is an easy target.

One major potential drawback to using a transceiver of any type as a physical monitoring device is the potential for alerting those you're monitoring with accidental traffic. In Afghanistan much of the chatter over the Taliban repeaters was monitored with captured radios. One of our partner teams had one during a planned surveillance mission that was to conclude with interdicting a Taliban leader. They were watching his safe house, and even though they had done everything else right, one of the guys on the team accidentally keyed up their radio during one of the Taliban transmissions. To make matters worse, he didn't know he had done it, and carried on a low conversation all of which was now transmitted to the Taliban. So now that they knew Americans were in the area, they dispersed, and all that work went to waste.

Frequency Counters

A frequency counter is a simple device that lists the frequency it detects at any given time. The close call and F-tune features built into Uniden scanners and Alinco communications receivers, respectively, are a type of built in frequency counter that tunes the device to the frequency. The drawback to both of those, however, is the limited frequency range and necessary proximity to the signal to capture it. Most frequency counters today have a range far higher than communications receivers or scanners, enabling us to map other signals in our area of operations such as cell phone signals, surveillance bugs, WiFi, and possibly even drones.

Frequency counters also have another incredibly useful capability when it comes to exploitation. Frequency counters tell us the sub-audible tones that are being transmitted, if any, by the detected signal. These are known as **CTCSS** (*continuous tone coded squelch system*) which are standardized tones that, when enabled, prevent accidental reception of a signal. These are very common in Amateur ("ham") radio as repeater tones, **Family Radio Service** (FRS) and **General Mobile Radio Service** (GMRS) as 'privacy tones'. When used, these are a basic communications security method, preventing the radio from responding to attempted jamming. If the CTCSS is identified, however, this basic attempt at communications security is broken.

Spectrum Analyzer – The Game Changer

Spectrum analyzers are devices that monitor a broad space of the radio spectrum all at once, giving the user a visual spike where it is receiving the signal. This makes intercepting traffic extremely fast even though there is no audio output. Spectrum analyzers used to be limited to labs testing the efficiency of transmitters and to dedicated, fixed site signals intelligence platforms due to their size and expense. The modern era has made them shockingly inexpensive, however, and incredibly small in size. Even the cheapest spectrum analyzer should be considered to be a must-have piece of gear for a SIGINT team on the ground.

My own tactical SIGINT package these days consists of a TinySA Ultra and a communications receiver, along with the necessary antennas for signals intercept and radio direction finding (RDF). This package enables me to rapidly identify frequencies of interest, identify the type of transmission on the waterfall of the Spectrum Analyzer, then physically monitor & RDF them with my communications receiver. Its a quick and efficient package that meshes well with a larger tactical loadout in addition to be flexible and simple enough to be used basically anywhere.

TinySA Ultra

The best of the handheld spectrum analyzers on the market today is the TinSA Ultra. When the original, the TinySA, hit the market a couple years ago I thought it was a game changer due to the user's ability to visualize intercepted signal spikes and have a built-in waterfall to see the nature of the signals themselves. As neat as it was, it had a couple of serious drawbacks. The first was that the screen was only 2.5in in size, which made it difficult to see and use. The second issue was that it had one antenna port for 0-300 MHz and the other for 300-960MHz. So it was good for what it was, but there was substantial room for improvement.

Enter the TinySA Ultra. Double the screen size, limiting the antenna port to one outlet for 0 -1000 MHz and increasing the overall reception range to 5.3 GHz, allowing it to intercept everything from HF transmissions to cell phones and drones. While there is no audio output, the TinySA Ultra allows the user to rapidly identify signals (in the form of spikes on the display) and use the waterfall to see the image of the traffic itself. So once we see the images in the waterfall we can compare them to the SigIDwiki database to identify what they are even if our communications receivers cannot decode it. Not only that, but we can gain LOBs on the signals themselves with directional antennas. At the tactical level this makes the utilization of any type of encryption irrelevant – the persistent presence of the signal itself, once detected and identified as hostile, becomes the target.

As an aside, this was the case in Bakhmut, Ukraine. Russian forces, spearheaded by Wagner Group's prison conscripts, were reluctant to enter the meatgrinder that is urban conflict.

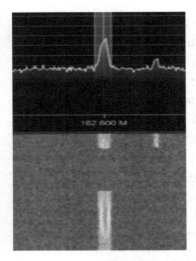

Instead they implemented the age-old Russian doctrine of heavy artillery bombardments ahead of the forward line of troops (FLOT). This time, however, they weren't blindly launching Katyuchas, it was munitions equipped with SIGINT sensors.

The target? VHF and UHF DMR signals which Ukrainian forces have been using since the outbreak of hostilities in Donbass in 2014. The flawed thinking, all of it rooted in the Global War on Terror's lack of an electronic warfare adversary, has led to incredibly sloppy discipline in tactical communications. So while DMR's greatest attribute – the ability to send text messages – was fatally underutilized, their longer duration voice traffic was immediately targeted with indirect fire. The same technology fielded in those weapons, at a basic level, are used in Spectrum Analyzers.

Getting started with the TinySA Ultra is fairly straightforward. The power switch it located on the top of the device beside the cursor wheel. The supplied antenna connects to the RF outlet – its an SMA Female – and while the antenna works well enough, its fragile and misses a lot of weak signals. We can do better. I run BNC adapters on all of my antenna ports. This makes running coax cables to larger, purpose built antennas much easier as well.

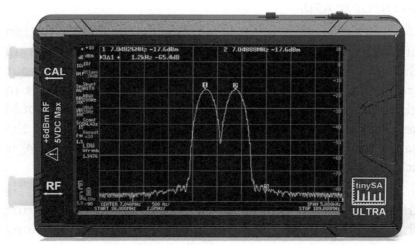

Once you switch the device on you'll notice the frequencies in yellow in the top corner of the screen – this is the display of what frequencies the device is picking up activity. The bottom left is the battery display and at the very bottom you'll see the START and STOP frequencies – this is the span that the TinySA is analyzing in real time.

Tap the screen to bring up the menu which will pop up on the right side. The device has several options for other functions of spectrum analyzers, such as tracing spurious emissions and transmitter quality, but since what we're concerned with here is configuring it for signals intelligence collection there's two options that we'll be using: FREQUENCY and DISPLAY.

Tapping FREQUENCY opens the menu to configure the bottom and top frequency range limits. So, for example, we know our target is running Baofeng radios which transmit from 136-174 MHz in VHF and 400-470 MHz in UHF. We'd be wasting our time analyzing a huge frequency space, so we set the limits to 136 for START and 174 for STOP. If we wanted to listen to UHF, we'd set START to 400 and STOP to 470.

The frequency is set by tapping the START button on the screen and this opens a new menu that takes up the whole screen and looks like a calculator. Enter your frequency number, the hit G if the target frequency is in gigahertz, M for megahertz, or K for kilohertz. You should see your new START frequency at the bottom left. Do the same for STOP and you've set the limits the TinySA is scanning in real time.

Wherever you see the spikes on the display is where the TinySA is detecting activity. The frequency its picking that activity up is displayed in the top left corner along with the signal strength in decibels. What we want to do is enable the waterfall as well, taking the same advantage of an SDR and putting it in your hand. Tap the screen again to open the menu and press DISPLAY. From there, tap WATERFALL. The bottom third of the screen now becomes a waterfall display showing the image of the signals you're intercepting. You can double this in size by tapping WATERFALL again. This is how I run mine. The other advantage to having a large waterfall display is the intercepted signals will remain on the display for a long period of time, so I can go back to a specific frequency I intercepted. You can control by rolling the cursor wheel left and right, which moves the flag icon back and forth, displaying the frequencies its landing on.

The TinySA Ultra looks complicated at first glace

but don't let that fool you. Its a simple piece of gear that is, without a doubt, the real game changer in signals intelligence on the ground. Older and much more expensive equipment, such as the RF Explorer, could be used similarly to the TinySA, I've found that this is a far more intuitive tool at less than half the price. And as a testament to its capability, students in the Signals Intelligence Course used to spend much time targeting the electronic opposing force with conventional scanners and communications receivers. The TinySA cuts that reactionary gap down considerably. Adding to that is the device's capability to detect drone and WiFi RF emissions (in the GHz range), this is a must-have portable package in a non-permissive environment.

Software Defined Radio (SDR)

SDR is, without a doubt, the most revolutionary piece of equipment in the SIGINT trade. At its most basic level, it takes a small USB dongle configured to receive signals and when paired with many open source software suites, can be used for both spectrum analysis and monitoring tasks. Further, using an SDR with specific types of software grants the operator the ability to exploit multiple types of digital signals, from two way radio to skimming cell phone data, with even a basic setup.

A visual display, known as the waterfall, gives us a visualization of the received signal. This is important for identifying specific types of traffic, especially digital signals. Examples of these signals are contained on SigIDwiki [sigidwiki.com] as discussed above. These come in handy when conducting radio pattern and traffic analysis. SDR in civilian hands has been used for everything from watching local broadcast TV and listening to commercial radio to tracking aircraft transponders using a software package called airspy. Some of the more interesting applications include creating a IMSI (*International Mobile Subscriber Identification*) skimmer for catching cell phone traffic in a given area and a platform for emulating digital signals (*such as P25*) in higher end devices, such as those from Apache Labs, for exploiting digital radio.

SDR is best suited for utilization in two roles: either as a fixed position listening post (LP) or mobile inside a vehicle. As powerful a tool as SDR can be, it is not a system that lends itself well to tactical application in the field. The system requirements demand a processor beyond what mobile devices can handle and the USB connection is not rugged enough to withstand field use. Much of the advantage to SDR, the waterfall image, is found in the TinySA Ultra spectrum analyzer. Inside a guerrilla base, safe house, or LP site, however, it is a major force mutiplier.

Release the Kraken! Kraken SDR

One of the more interesting developments of SDR technology is the Kraken SDR. The Kraken is not simply another SDR, but serves as a radio direction finding (RDF) device when paired with a specific type of antenna known as an Adcock array. This creates a tool not just for interception but also to DF the source of RF energy, whether that emission is on the ground or in the air. It has been pressed into service in Ukraine (and likely elsewhere) as a passive radar system, meaning receive only, to detect incoming aircraft including drones. This level of sophistication at a relatively low cost has not been previously available to civilians and is one that should become a priority in obtaining for any guerrilla unit, at least while the opportunity presents itself. That window may be quickly closing.

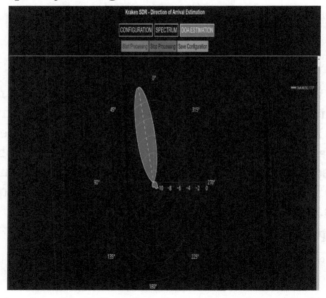

During the fall and early winter of 2022, Ukrainian defense forces from both the Azov Battalion and elements of the International Brigade put several of the Kraken SDR units to use in the Kherson region. This allowed them to rapidly detect the presence of Russian ground forces and air assets. This would later be used to detect the presence of Iranian made drones being fielded by the Russian military in its attacks on Kyiv. The streamlining of signals intelligence recognition, collection and then targeting for countermeasures was critical in the defense of the areas in conflict. Because of this, the Kracken as of this writing may become limited in its availability to civilians.

While the KrakenSDR comes with a decent mobile antenna array, there's much better ones on the market. Arrow Antennas out of Arizona builds one of the best and its what I run in my semi-permanent Listening Post (LP). The reception and direction finding quality is superior due to its more specific attenuation to specific frequency ranges. More on this in a bit.

PortaPack H2

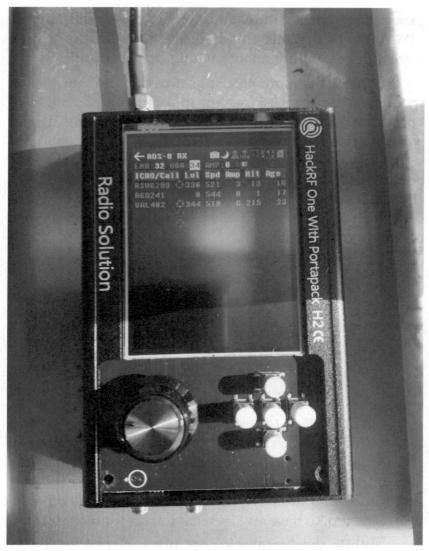

One of the more interesting devices coming out in the past few years is the Portapack H2. It takes a number of the advantages of SDR and puts them in a portable package. In addition it has an included transmitter that's useful for replaying (spoofing) certain captured signals, ranging from garage door openers and car keyfobs to GPS and aircraft ADS-B beacons. While this is illegal to do in the United States, it does present an interesting set of capabilities.

One of the more potent uses of the PortaPack H2 is spoofing a target's GPS system in a given area. Since we know that most of the more advanced government-fielded weapon systems are guided by GPS, the ability to spoof that signal may create just enough of a reactionary gap that can be exploited. On the note of spoofing, the PortaPack H2 can also replicate ADS-B signals, leading to confusion of where certain air assets may actually be located. This also applies to commercial drones.

In its ADS-B receiver mode it generates a list and location of aircraft signals its intercepting in real time. It also has the capability of displaying a map (that you load into the memory) which creates a portable aircraft geolocation monitor. If you're tracking aircraft off-grid, this is the tool to have and its what I use it for most often.

While the PortaPack does not replace the TinySA Ultra in my own loadout, it does have some interesting applications based on what it can do. Although its still fairly clunky to use in many of its intended roles it is worth having, if for no other reason, than to be an awareness tool.

SDR Software

There are literally too many SDR software suites to mention in this volume, with more being added daily. They are usually however open source and normally free to use, ranging from intuitive for the end user to requiring substantial knowledge in computers and signals collection in order to use. One of the simplest software packages for the end user is **AirSpy**. Lacking a high degree of sophistication, AirSpy enables the user to capture the basic image of a signal on the waterfall paired with the audio output.

A more advanced program specifically suited to digital signal exploitation is **DSD+**. DSD, or Digital Speech Decoder, is a software suit designed to identify the properties of a digital signal in an effort to either decode, possibly decrypt (*depending on the level of encryption*) and fingerprint the specific signal properties of the target. This is a very powerful tool when conducting traffic analysis of a target which has created a specific pattern of communications. It has been used to fingerprint and map the hierarchy of militant groups operating in the Middle East utilizing DMR for well over a decade. The same can be done for literally any other type of digital waveform based on the metadata it contains.

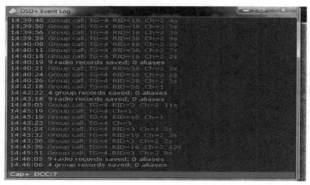

In addition to the programs that are in the public domain, there's a couple of dedicated operating systems (OS) as well. One of the benefits of running a dedicated OS with an SDR is that you're using all of the computer's processor power for SigInt tasks and the tools that are embedded can end up overclocking many. One of the older ones that I'm familiar with is SigintOS, which has since fallen off the map in terms of follow-on support. Legacy distros are still floating around however and it was designed to run with BladeRF to be used as a potential exploitation tool.

The current best OS on the market is Dragon. It picks up where SigintOS left off, streamlining the ability to run any of the SDR devices in one purpose-built package. While neither are necessarily what I'd call a tactical use package, they do offer some advantages for fixed positions and dedicated listening posts in the form of having all of your tools streamlined into one platform.

CHAPTER 3. SIGINT ANTENNAS

The most critical component of any SIGINT collection package, as with communications in general, is the antenna. A proper antenna systems improves our reception range, attenuation, and ability to get a proper **Line Of Bearing** (LOB) on the source of a signal. Antennas for signals collection fall into one of two categories: **Omnidirectional** and **Directional**.

Discone omnidirectional receiving antenna

Omnidirectional Antennas receive equally well in all directions, giving a 360 degree reception range. These are used in a LP first for establishing a baseline for the active signals inside the area of operations and for situational awareness. If it transmits, it can be detected and handed off to a team tasked with gettting a bearing on the **Point of Origin** (POO) as discussed in the exploitation techniques chapter. One very common example of an omni-directional antenna is known as a **Discone**. The discone antenna is primarily a receiving antenna with ground plane elements of various lengths cut for optimum use on various frequency ranges. At its heart, the Discone antenna is incredibly similar to the Jungle Antenna for transmitting – and that same design works extremely well in this role.

Directional Antennas are those that optimally receive in a specific direction. This allows us to determine the strongest (*or in the case of a loop antenna, the weakest*) signal based on the signal strength indicator, giving us a **Line of Bearing** (LOB) of the signal itself. Any received signal, especially those on the ground, will have a LOB. There are three general types of directional antennas –

the Yagi, the Loop, and the Adcock Array. Each of these are used by SIGINT teams on the ground in an effort to integrate those assets for targeting. When done right, it can become extremely effective.

Yagi Antenna

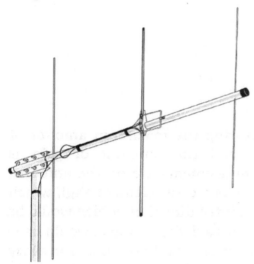

Yagi Antennas are probably the best known and most common directional antennas, since they used to be extremely common on top of people's houses for TV reception. In fact, if you were familiar with turning the antenna in the direction of the TV station to get the best reception, you're already very familiar with the concept of radio direction finding whether you knew it or not. Because in a nutshell, that's how you use a Yagi. Turn it until you get the strongest signal – that's the LOB for the origin of your target signal.

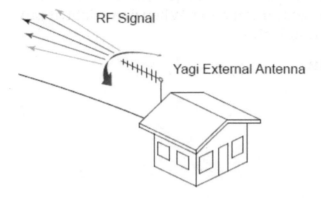

Yagis work by creating the maximum amount of gain in one direction, and that amount of gain is determined by how many elements are on the antenna. The antenna shown above is a three element Yagi, which has 7.5db of gain in the pointed direction (which would be to the right). A five element yagi, for example, would have 9.5db in the pointed direction. We know that for every 3db of gain, the signal strength doubles and this works in orders of magnitude (for each 3db of gain it doubles again). What that also means is that not only will it receive a signal that much stronger, it'll also give a tighter azimuth in that direction.

I explain Yagis by making the analogy to light. A Yagi is like a flashlight, with the illuminated direction from the flashlight being the direction of the radiated RF energy. The more elements, the tighter than beam of light will be. But in the case of signals intelligence that illuminated area is where you're receiving the RF energy of others the strongest. Another example which we used before was the old over the air TV stations – we'd turn the Yagi antenna in the direction of the transmitter for the strongest reception.

That's what you're doing here.

Loop Antenna

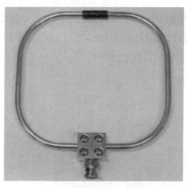

Loop Antennas work the opposite. They receive equally well in most directions save for the center, where they receive nothing. That is known as a null – where no signals are heard – and this gives a LOB to the source of the signal itself. Loop antennas have a couple of advantages over Yagis. The first one is the size itself – they're much smaller. Smaller is usually better in terms of equipment footprint. The other advantage is that, unlike Yagis which usually are still receiving signals in most directions, finding that Null, where the signal is weakest or not heard – is easier to do in an environment where the signal is either extremely strong or there's lots of signals on nearby frequencies. A loop generally will give you a very tight azimuth for the LOB, whereas a Yagi has between a 60-80 degree of variability on the sides of the antenna. A Yagi is better for consistent monitoring (because you're still listening to the traffic) while a loop is better for directing your movement (due to the tighter azimuth).

Using the loop is pretty simple. The one pictured above comes from Arrow Antennas and has a BNC connector as its base. You can attach this directly to the top of your handheld radio. It swivels at the base, so you can hold the radio in one hand and rotate the antenna with the other holding the black insulated portion at the top. Wherever you receive the weakest (or, ideally, no signal) is your null. Look through it like you would the rear sight on a rifle, and that's your direction.

We've been using Loop antennas for a long time. One of the more legendary stories about Ernest Hemingway was that his yacht, the Pilar, was equipped with a loop antenna on a HF radio that they used to call Huff-Duff units (HF-DF). He'd use these to try to gain LOBs to German U boats operating in the Carribean when they'd surface to send their communications reports (via HF) and encrypted via the Enigma. Armed with a Tommy gun, a case of grenades and probably several cases of Bacardi, he was never successful but he also wasn't as full of it as many historians like to joke about. There was a very real science to what he was doing and given his history I don't doubt that he was serious about hunting the Nazis.

One of the best places to source DF Loops is through Arrow Antennas out of Arizona. They build a high quality, robust piece of equipment that works extremely well as part of a DF package. I've used one for a number of years in class and have found it to stand up to an impressive level of abuse.

Adcock Antenna Array

The **Adcock Array** is a group of omnidirectional antennas making a circle all connected together. The antenna which receives the strongest of the signals gives the LOB. The Adcock was invented in the early days of WWII by the British in order to detect German U boats and incoming aircraft. The design was so successful that it became an inexpensive alternative to radar for small airports in the post-war era and is still commonly used today in Lo-Jack systems, the Wolfhound radio direction finding unit we fielded in Afghanistan, and currently with the KrackenSDR system.

An Adcock is simply a series of dipole antennas spread out with each connected to a discriminator. What this does is indicate which antenna is receiving the strongest of the signals and gives the user a line of bearing based on that. While they can be used on the move and sometimes can be employed very effectively (depending on the target), the Adcock Array works best when used in a fixed position, otherwise known as a **Listening Post** (LP).

A Word On Antenna Theory

In *The Guerrilla's Guide to the Baofeng Radio* we explore antenna theory in terms of optimization for transmitting. Concepts such as antenna gain and standing wave ratio (SWR) apply directly to the radio's ability to efficiently transmit. But this also has an important role on the radio's ability to receive. The closer the antenna is to an electrical match, meaning the lower the SWR, the stronger it will receive those signals. Further, the more gain an antenna has, especially in a specific direction, the tighter the LOB will be to the source of the identified signal. For example, a VHF signal is best received by an antenna cut for VHF. A UHF signal is the same. While one can certainly receive the other, and in some cases may do so very well, the best results are going to be had with a purpose built antenna.

Why should we worry about matching antennas to the frequency range we're trying to receive? If we're not transmitting, does it really matter? The answer is yes – it absolutely does. When transmitting is the goal, Standing Wave Ratio (SWR) is important. We want as close a matched antenna as possible for transmission efficiency. With signals interception, however, that same efficiency is critical to peak reception. In a nutshell, the antenna receives the best when its closely matched to the signals you're trying to receive. For example, a VHF antenna is going to receive VHF signals better than a UHF antenna would. The opposite of this is introducing **Attenuation**, or, a reduction in the signal to further refine the LOB of a nearby strong signal. Either way, we must use the antenna for the matching frequency range for the best performance first and for that reason we need to understand the basics of antenna construction.

Antenna theory is begins with understanding that

all antennas are a form of a dipole. When we add more components to an antenna, known as elements, we get changes in the antennas effects. Sometimes this is for improving the reception in an omnidirectional way, other times it is for improving the signal reception in a specific direction. These are all based around the placement of the elements themselves. Since a guerrilla may not be able to source manufactured antennas in most cases, the ability to build your own equipment becomes vital.

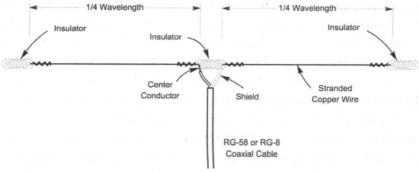

Antennas have two halves – a positive and a negative element. These form what's known as a dipole antenna which is a half wavelength long. The calculation for a dipole antenna, end to end, is:

468 / Frequency in MHZ = Length in Feet

Since on a dipole there's two halves (which are a quarter wavelength long), we split 468 in half to get 234. Our formula for each piece of wire to cut now becomes:

234 / Frequency in MHZ = Antenna Element (the wire you're cutting) Length in Feet

Every guerrilla group tasked with either communications or SIGINT must have a number of raw materials at their disposal. The first one is wire itself. Any type of wire can be used as a radiating element, even copper welding rods in the case of the Yagi. Another required item is a split post BNC adapter (also known as a Cobra Head) which is used as a the feed point for the antenna. Last, you'll need coax cable to connect the antenna to the radio – I prefer RG-8X with BNC connectors. Its cheap, simple and robust.

Field Expedient Antennas

A receiver, even the most high quality, most sensitive ones on the market, are nothing without a purpose built antenna and this is why I spend so much time talking about them. And while we could spend a bunch of money on factory made equipment, the reality is that you may just have to rely on the equipment you know how to build based on materials you can source in a given working environment. The reality of guerrilla warfare is, after all, that fighters have to make do with what they have. The great news is though that often, believe it or not, home built of field expedient antennas (when built right) perform every bit as well if not better than their factory made counterparts. Why? They're cut for a specific frequency or a more precise average of a frequency operating range, meaning the users tailor make them for purpose.

The two easiest antennas to build for SIGINT are the Jungle Antenna and the Yagi. The former is for scooping up everything that transmits – more on this later – and the Yagi for both consistent monitoring and gaining a Line of Bearing to the source of a signal. In both cases we can build these antennas from parts found in almost any hardware store, save for the Cobra Head, or Split Post BNC Adapter. But even if you don't have one of those, cutting one end of coax and soldering the center (positive / hot side) and the insulation braid (negative / cold side) works fine in a pinch. On that note, if you're running coax cable for simply listening, RG6, which is the most common TV coax cable, works great. Its 75 Ohm, which means the transmission quality on reception has low loss. If you're going to be transmitting, RG-58, 8X, LMR, etc. is preferred because 50 Ohm coax is what transmitters are designed to handle.

Constructing a Homebrew Jungle Antenna or Discone

The first, simplest antenna to build is the Jungle antenna for omnidirectional reception. The Jungle antenna is not much more than a dipole with extra elements added to the groundplane side. What this does is create a radiating and receiving pattern that's equal in all directions and this is the first antenna we need to build for scooping up all of the signals in our working environment. This makes it the first go-to antenna. In *The Guerrilla's Guide to the Baofeng Radio* I cover building it as your first transmitting antenna for both its simplicity and its gain (6dB). Obviously for a general purpose transmitting antenna this works very well, but as a receiving antenna it becomes a discone as described above.

All of my field expedient antennas begin with split post BNC adapters, also known as a Cobra Head. These make building antennas simple and robust. You'll notice a red side and black side – this is for positive and negative, respectively. The positive side, or hot side, has one wire.

The negative side, or cold side, has at least three wires. The wires are cut to length for the targeted frequencies and they are each a quarter wavelength long (234 / Frequency = Length in Feet). For example – if we want to listen for VHF signals, say, 136-174mHz, we want to cut them for the average of that frequency range. The next step is simple; attach one on the red and three on the black. The three on the black creates your groundplane and we'll be attaching each of these to spreaders. It creates what looks like a pyramid and receives best in the range its cut for. But this doesn't mean the antenna won't receive in other ranges, it just means its not optimum.

If you don't have time for the math, here's a quick reference of the most common traffic frequency spaces and their uses for targeting purposes:

Quarter Wave Antenna Length Quick Reference:

- 160m (1.8-2 MHz): 128 Ft
- 80m (3.5-4 MHz): 64 Ft
- 40m (7-7.3 MHz): 32 Ft
- 20m (14-14.350 MHz): 16 Ft
- Citizen's Band [CB Radio] (26-28 MHz): 108 Inches
- 10m (28-29.7 MHz): 8 Ft
- Military Ground (30-88 MHz): 4.5 Ft
- 6m (50-54 MHz): 4.5 Ft
- Broadcast FM (87-108 MHz): 2.5 Ft
- Civil Air Band (108-137 MHz): 1.95 Ft
- 136-174 MHz: 19 Inches
  - Marine VHF, Railroads, Amateur 2M, MURS, Public Safety VHF

- 200-380 MHz: 9.36 Inches
  ◦ Amateur 1.25M, UPS servicing, US TACSAT, Russian SATCOM
- 400-512 MHz: 6 Inches
  ◦ Amateur 70 CM, GMRS / FRS, Public Safety UHF
- 902-928 MHz: 3 Inches
  ◦ Mesh networking devices, GoTennaX, XTS radios
- WiFi and Drones:
  ◦ 2.4 GHz: 1.17 Inches
  ◦ 5 – 5.8 GHz: .52 Inches

Constructing a Yagi Antenna

As discussed above, the Yagi is the primary tool for getting a Line of Bearing (LOB) to a signal. I covered the construction in step-by-step detail in *The Guerrilla's Guide to the Baofeng Radio*. The most common one and the first one you should build is the three element Yagi as its both compact for field use and broad enough in reception angle to get a quick LOB to a targeted signal. The more elements you add, the tighter that reception angle becomes and the higher the gain on the antenna. Higher gain equates receiving a signal over a longer distance that you possibly wouldn't detect or have the ability to monitor.

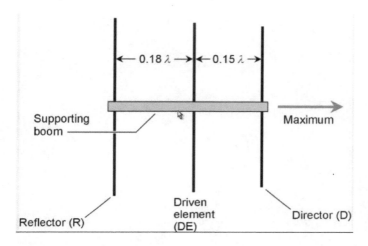

Yagi antennas look complicated, but in practice they're actually extremely simple to build. When I start I lay the center, or the boom, down on a table and begin the measurements for the spacing there before I make any other measurement or cut any wire. Since the antenna works based on the spacing between the elements, I always begin with this step first. The center, or the Driven Element (DE), is the only part that's actually connected to the radio. This part is a Dipole. The other two parts, the Reflector (R) and the Director (D), are spaced off of the center. That spacing labeled above is measured in terms of wavelength, with the constant being 936. So the constant for the space between R and the DE is .18 wavelength. If we multiply 936 by .18, we get 168.48. So how do we find the actual distance?

Reflector to Driven Element: 168.48 / Frequency = Length in Feet

Let's do the same for .15, which will give us 140.4.

Driven Element to Director: 140.4 / Frequency = Length in Feet

Mark those lengths on the boom and now we have the template for building our Yagi. Next we're going to calculate how much wire we're cutting. A quick note about wire – anything that conducts works fine, but for Yagis we want something that's fairly rigid. One of the better field expedient materials we've used in class is welding rods, although tape measures and coat hangers work well too.

When I start, I cut the Reflector first since its simply one continuous piece of material. Its slightly longer than the driven element:

510 / Frequency = Length in Feet

Next I cut the Director since, like the Reflector, its

one piece of material also. Its slightly shorter than the Driven Element:

425 / Frequency = Length in Feet

At this point you can attach both of these to the boom. We've done everything from super nice elaborate ways of attaching them to just using electrical or duck tape to keep them in place. Both work just fine even if it ain't pretty. With that said, larger Yagi antennas (with more elements) simply add more directors. What this means is that if I want to build a 5 element Yagi, I simply space out two more Directors on the boom and cut them the same length. It really is that simple.

Now its time to cut our Driven Element, or DE, which is connected to the radio itself. Its nothing more than a dipole, and we calculate this with the formula:

468 / Frequency = Length in Feet

We're going to take this piece of wire and cut it in half – each half will be connected to the Cobra Head as the dipole center. Hook that up to a run of coax, attach it to your radio, and you're good to go. Congrats, you just built a Yagi for direction finding!

Antenna Polarization

Signals are polarized when they transmit off of an antenna and this is based on the orientation of the antenna itself. **Antenna Polarization** then becomes another critical point of interest. If the antenna is vertical, a signal is **Vertically Polarized**. Parallel with the ground, **Horizontally Polarized**. If transmitted from a loop, it is **Circularly Polarized**. There is a 12db difference in signal strength from one polarization to another. What this means in simple terms is that if you're receiving a signal that's vertically polarized on a horizontal antenna, it may be extremely weak. Match the polarization and you'll receive the signal stronger.

How to know? Its simply based on averages. The bulk of signals being collected in a tactical sense are vertically polarized; they're being transmitted from an antenna that's vertical. This would include handhelds, vehicle mounted and fixed position equipment. But with that said understanding polarization is important to know not just for the best reception but, on the other hand, introducing loss should you have nearby strong signals.

If a targeted signal is strong enough and close enough it may very well overpower the reception capability of the antenna and device you're using to direction find. When this happens there's no direction you point a Yagi that will diminish the signal or no point to aim the loop to find a null. So what we do is add attenuation through changing the polarization of the antenna itself. By introducing loss we can achieve a better line of bearing to the strongest signal when using the Yagi.

Groundwaves – the real target of interest

Radio waves in and of themselves propagate, or travel, two ways: **Groundwave** and **Skywave**. As the name implies, groundwaves travel along the ground and skywave reflects off the ionosphere. Groundwaves are most commonly understood as Line of Sight, or LOS. This is often attributed only to the HF portion of the radio spectrum but this is not the case. Its simply due to the poor reflectivity of the layers of the ionosphere getting into the higher VHF frequencies and above. But with that said, as with antenna polarization, ground waves become the target for direction finding with our antennas even in the HF frequency range. As with the communications realm, the higher in antenna height we go the more line of sight we have. Higher is better for receiving as well as transmitting if maximum line of sight is the goal.

Obtaining a Line of Bearing on groundwaves can be tough, depending on the terrain. The more terrain features that are present in a working environment will scatter a signal. For the SIGINT collector this may only give you a LOB to the point where the signal is being reflected or bounced. Think of yourself in a dark room and a flashlight shines in a mirror; the direction the light bounces is a the same concept.

In the mountains communicating in the higher end of HF or low band VHF frequencies its difficult more often that not to gain a solid line of bearing to a transmitter. Higher frequencies, UHF two way radio and especially mesh networking devices, make this incredibly easy due to their poor reflectivity. Higher frequencies penetrate terrain features and buildings rather than reflect. For this reason in the later years of the Afghan war the Taliban made use of CB band (26-27 MHz) radios. It wasn't that they couldn't be detected or heard, they knew that much. But the frequency range itself and the given terrain negated most of the ability to be targeted via their transmissions while keeping in mind that they were being consistently monitored. Moral of the story? The ends justify the means. Low tech wins again.

CHAPTER 4. SIGNALS INTELLIGENCE COLLECTION AND TARGETING

The first task for any signals intelligence team when standing up collection in a given area is to establish that Baseline as described in Chapter 1. Once the baseline is recognized, a team can quickly conduct multiple tasks: filtering what is of intelligence value versus what is simply a pattern of life, identifying anomalies that may have tactical value, and focusing attention on what is actionable versus what is not. How do we define what's actionable? Simple answer - what can I do with the data I'm collecting.

Perfect example – a SIGINT team on the ground may receive an encoded and possibly encrypted signal in the VHF portion of the spectrum. Not on any of the previously identified frequencies in use, and not in a mode that anyone recognizes. Looking at the signal on the waterfall, it looks similar to Digital Mobile Radio (DMR). The signal is strong enough to get a bearing, but is moving slowly. It is reasonable to suspect this is a team using DMR handhelds as a tactical radio. We know that the range of their radios are not more than two miles based on the terrain, the consistency of the signals and the strength of the signal being received, and the apparent lack of radio discipline. It does not matter at this point what they are saying – the fact we are receiving them in real time, understanding these knowns allows us to effectively target them. We could, given time, dedicate assets to decoding and possibly attempting to decrypt the signal. But why would we? If we know roughly where they are and their direction of travel, sending a team to get eyes on them is a more efficient use of our resources in real time, as with the example in Chapter 1 of targeting ISIS' DMR.

In that example we rapidly compared knowns versus unknowns for a place. The signals did not match anything we had previously intercepted, and since it was new, must be assumed to be hostile until proven otherwise. Had our SIGINT team not done their prior work in establishing a baseline, we never would have known. The electronic spectrum may be broad, but it is not impossible to understand. It does however require dedicated personnel assigned to the lone task of SIGINT in order to be successful.

Categorizing Intelligence Targets

When we conceptualize SIGINT as a task, most fail to understand the specific categories of intelligence targets themselves. SIGINT is best understood as the other side of the communications coin, every bit as important. For that reason, before we can begin F3EAD, we must first recognize the category of our intended target. You must know what you are hunting before the hunt can begin. Communications fall into one of three mutually exclusive categories:

1. **Sustainment Communications**
2. **Tactical Communications**
3. **Clandestine Communications**

Sustainment Communications are those that provide stability for an area. These include law enforcement, emergency services, public utilities, civil aviation and amateur (ham) radio. These generate the most amount of traffic. For signals intelligence purposes, these are normally used for situational awareness and establishing a baseline of activity for the operating area.

Typically, sustainment communications are what most focus on monitoring first. As covered in Chapter 2, this is where the basic scanner fits into the mix. While it is important for awareness purposes to monitor the local infrastructure, this does not usually provide us with anything actionable. It may, however, give us updates on second and third order effects of whatever activities are underway in that given area. These usually are better accomplished through other means.

Tactical Communications are those utilized for coordinating fire and movement during an operation. Tactical communications are what most think of when they think of implementing radios into small unit tactics. These are the most important targets of signals intelligence and will comprise the bulk of signals intelligence collection at the small unit level. Doing so provides actionable intelligence in the form of an early warning, geolocation data of the source of the signals, and if possible, mapping the hierarchy of the organization.

Tactical communications are the most misunderstood topic of small unit tactics. Frequently amateurs and armchair warriors place a premium on command and control at all times, leading to major issues with micromanagement among conventional forces (*and even those claiming to be unconventional*). This leads to a huge amount of radio traffic being transmitted during movements. While a premium is placed on encryption of the signal itself, nothing is done to mitigate the source or sources of the signal. Just like with our example above, the more signals, the larger and more defined the pattern of life, the easier it is to track and exploit. The bulk of the techniques in this book are intended for tactical signals exploitation (TSE).

Clandestine Communications are those which coordinate activities over a much larger region, possibly outside the area of operations itself. These have the utmost level of encryption and contain little, if any intelligence value if intercepted. In *The Guerrilla's Guide to the Baofeng Radio* I address the techniques of burst transmissions and digital operations as a means to accomplish this. The same techniques can be applied over a broad array of platforms, from HF radio to email and social media. While many resources could be dedicated to breaking a clandestine transmission once intercepted, understand that for a guerrilla force, or even most government-backed intelligence organizations, its most likely a waste of time absent prior knowledge of the sender and a short list of likely recipients. If the sender and the recipient have a level of training and coordination to send traffic in this manner, and that's assuming its not false traffic to start, then whatever instructions are contained are very likely to already be underway. The intelligence value then is knowing that the message was sent, that the point of origin maintains some level of control, and now we must rely on

HUMINT to map association between potential recipients. It sounds complicated and it is. This is why a premium on clandestine communications techniques needs to be placed on a guerrilla force.

Creating a Signals Baseline

The first order of business is creating a baseline for all of the signals emitted in the given area of operations. Hunting for signals, their sources, and the purpose behind them is a completely fruitless endeavor if one cannot differentiate what's routine from the abnormal. Fortunately there are a number of resources to make this simple at the outset. Depending on your area of operations, the first step is indexing the open source signals references themselves. In Chapter 2 I listed Radio Reference as a primary means of sourcing that baseline data for a given location. It contains all of the licensed data for a given area and, in many cases, the geolocation of the source of the signals as well.

For example – we have an airport nearby. The international civil airband is 108-137mHz. Intercepting signals in this range are either going to be the aircraft themselves or the airports handing their radio traffic. Airports have fixed, published positions. So let's say we're receiving unknown traffic in that frequency range and direction find (*DF*) the signal. More than likely that line of bearing we obtain is going to point to the local airport.

We have to establish a baseline for the patterns of life of a place to rule out what's wasting our time. Frequently in the Signals Intelligence Course I have students that intercept routine dispatcher traffic thinking its of value when in fact its normal for the area. They're wasting their time while the tactical communications of the actual target continue to go unnoticed. But we eliminate those through the process of first conducting a **Spectrum Search**, then a **Point Search**, while comparing the data to the open source databases (*Radio Reference and Sig ID Wiki*) to determine their purpose.

Spectrum Search

The first step in establishing that baseline is determining everything that's transmitting in our given area. From here we know what's normal for an area and what's not. Why would we want to do this? For starters, it tells us what we can eliminate from having to constantly monitor.

When getting started collecting signals its common to get false positives. You intercept something, it sounds interesting, and bam – you're dedicating time to monitoring it only to find out its a water meter at the local pump house. And that's fine, you didn't know and now you do. But the purpose behind our spectrum search is to find literally everything emitting a signal. Everything means everything – every blip on the radar, every tone that breaks squelch, everything. We're logging it, taking images of the waterfall, and at a minimum being aware of its presence.

This process takes a large amount of time. Its not unusual for the team tasked with Spectrum Searching to be doing this for hours and hours on end. Its a process that used to take a tremendous amount of time with scanners and communications receivers alone, but the invention of SDR and now the portability of the TinySA and devices like the PortaPack H2 have revolutionized the signals intelligence world. If I can see entire swaths of the spectrum at once, I'm watching everything that's transmitting in real time. Even still, expect this part to take up the most amount of the schedule. It might be boring but its the first major step.

From here we're going to take that data we've collected and compare it to known data sources. This is where Radio Reference and SigIDwiki come in. The more detail we can add into what we've collected, the better. Specifically what we're looking for is establishing what the purpose behind the signals we've intercepted actually are. From there we can decide whether its worth our time to continue monitoring or to write off as "good to know".

The other purpose behind the Spectrum Search is learning those areas that do not, in fact, have much traffic or are an otherwise a benign purpose. From the communications side this is what we'll be using to create our own communications plan inside a non-permissive environment. Logically, we could avoid the common traffic, such as public safety, which would obviously get us detected. But that said, we can also hide in the noise of another signal that serves a relative low importance, such as that aforementioned water meter transmitting in VHF. This technique is called piggybacking, and you have to be careful with it as to not get noticed or interfere with otherwise benign communications. But with that said, from the Signals Intelligence side, know that your query may very well be doing this right under your nose. The only way to know in both cases is conducting that spectrum search as a starting point.

CB radio is very, very good example of this. In the US there's tons of erroneous traffic – its literally the wild west of radio with very little enforcement as long as one is not causing too much trouble too frequently. And even if you are, its tough to get caught unless you're incredibly stupid.

For the potential guerrilla, this presents an interesting option. Knowing that CB is most likely to be ignored by basically everyone, has great coverage in nearly every environment, and has a huge amount of equipment out there for it, an underground group running scheduled data bursts over the CB band could be highly effective and very likely go completely unnoticed. It would be very difficult to direction find the source of that signal if its hiding within so many other erroneous signals, and that is once it was even detected. Just something to consider – everything that's old becomes new again.

Point Search

Now that we've conducted our spectrum search we've established what's transmitting, on what frequencies, and hopefully, why its transmitting. A point search is simply narrowing that focus down and limiting it to those areas that have the most traffic or are most likely to become targets. For example, if my adversaries were spotted with Baofeng radios for inter-team communications, it wouldn't do me much good to try to monitor them in the 300 MHz frequency range, because that's outside the range their equipment is capable of operating. But we do however know, based on the equipment limitations, they'll be operating between 136-174 MHz and 400-470 MHz. So our point search now becomes focused on those two frequency ranges.

With a point search we want to use both the spectrum analyzer and a communications receiver. The spectrum analyzer is displaying the signals in real time. Once we see the spike or the stream on the waterfall we tune the communications receiver to the specific frequency we've intercepted. Now we're hearing whatever the audio output actually is. From here we want to record that traffic for future analysis in order to exploit it.

One example of point searches from a long time back was an exercise I observed just prior to leaving the Army. A conventional Armored Brigade was on a month long exercise simulating combined arms maneuver against a peered adversary. What they didn't know is that this included an electronic warfare component from the NSA acting as OPFOR. They knew, based on published data, what frequency range the PRC-119 radio operates in...30-88 MHz. So even though it frequency hops every three seconds in that span, which is there to prevent jamming *(not for communications security as commonly believed)* it is far, far from undetectable. Quite the opposite. Their hops were watched and recorded and from there were emulated in a 'captured' radio. The Armored Brigade's communications were compromised rapidly and they never even knew it – more concerning, they did nothing about it until they were being jammed to the point of confusion at all subordinate units. Not good.

But understand that applies to you as well. They had set a distinctive pattern which, even though it was based on equipment limitations, did not even consider modifying in practice to avoid detection. Twenty years of fighting an unsophisticated adversary will do that, along with an unhealthy obsession for micromanagement. You may have equipment limitations but not regularly rotating through your communications planning is a sure-fire way to get compromised and exploited. And with that said, the more limited your equipment is, the easier it will be to intercept. If, for example, your query is observed using surplus police radios, like, say a Motorola Astro Sabre, I know its limited to a very narrow frequency range *(normally 150-161 MHz or 460-470 MHz although this can be modified)* and thus any communications security measures they may have in encrypting the signal becomes irrelevant – through my point search I found them through their very likely sloppy communications practices *(ie, not shutting up)* and can simply exploit them at the tactical level with my rifle. It is that simple.

Point Search Targeting and Surveillance

The purpose behind the point search is fact finding. We now know, or at least we think we know, what the operating frequencies of our target are. We've graduated past the Find and are now squarely in the Fix. Point Searches are all about surveillance.

First we want to establish what the nature of our target actually may be; what is their role in a larger strategy? Are they actually hostile? If they're part of an occupation force, what is their role within that occupation? What is their level of capability and / or familiarity with their equipment? What is their level of morale? What patterns are they establishing for themselves? How might exploiting one flaw cascade into other problems?

This is where the SOI covered in Chapter 1 comes in handy. As I'm surveilling a target I'm writing down first their frequencies, the operating mode and the callsigns. Those callsigns, by the way, *do not have to necessarily be spoken or consciously used by the target himself as an identifier*. As I've discussed earlier, radios with built in digital protocols have identifiers on a net that the users may or may not be aware of. DMR, P25 and NXDN are very good examples. This is due in part to rapidly identifying unauthorized users entering a net *(the master radio used by net control can identify these)* as a communications security measure but its also an inherent way for a Signals Intelligence team to collect **Positive Identification**, or **PID**, on specific targets of interest.

Going back to the ISIS example, this is partly how Omar al-Shishani was eventually identified on their DMR repeater network and eventually taken out. Shishani was a member of the Georgian Special Reconnaissance Group prior to becoming an Islamic militant. Leading up to the Russian invasion of Georgia in 2008, US Special Forces had advised and trained the Georgians for stay behind activities in the event of that eventual Russian invasion which had been signaled since at least 2004. On the communications front a primary focus was placed on tactical digital communications security methods over physical ones, and DMR repeater networks became the standard which the Georgians were trained up on. Afterwards Shishani would take those lessons and apply them to the formation of ISIS as their military commander. He did what he was trained to do, and despite those early successes, he was not completely aware of the pitfalls, specifically the metadata being transmitted despite the presence of physical encryption of the transmission itself. In large part it got him killed. Metadata strikes again. Its not what's being said, in this case, it was who's saying it.

But to the point of surveillance during the point search, we're looking specifically for hierarchy. Who's in charge, who after that, and so on and so on all the way down the line. During the Finish and Exploit phases this will become extremely important to monitor for potential action indicators or changes in target behavior, electronically or otherwise. Only targeting low-level players may very well be a waste of our time, but on the contrary isolating the mid-level leadership from the low level players normally cripples an organization.

Recognizing this hierarchy is determined by the patterns of life of the group you're observing. When are specific callsigns communicating can give you an indication as to the purpose behind it and their specific role in that organization. For example, let's say you're watching an enemy camp serving as a logistics hub for a larger operating area. They have to guard it, and those guards operate in shifts. This means they'll have fixed guard tower positions including the access points. The guard shift will have at least one rover, and he's usually the senior person there. But he answers to someone as well. I'd want to know is what his pattern of life is along with how often he transmits. The tower guards themselves are mostly irrelevant but the rover is the one to isolate, along with whoever is in charge of him. But you won't know this without first identifying those callsigns and/or radio IDs. Just a thought.

The last piece of making an assessment of a target, after determining hierarchy and purpose, is determining the relative level of training and discipline your target actually has. Let's say your targeted group is wearing matching outfits, has shiny, new, high end equipment, and at least looks the part of a serious threat. And they may very well be save for one factor – over reliance on communicating at the tactical level, ie talking way too much. This in an of itself becomes a point of exploitation. An undisciplined target, even if they look the part, will chatter non-stop. A disciplined target is one that communicates only when absolutely necessary and even then keeps it to extremely short (*3 seconds*) bursts.

    As an aside this is one of the major problems with the contemporary American shooting / tactical culture. Everyone wants to look the part, forgetting that what they're really doing is emulating conventional Infantry tactics with nicer equipment sold to them at a premium while conveniently dismissing the leviathan of a supply line, planning, and strategic objective it took to get to the firing line. It in no way is unconventional nor has it achieved much in the way of warfare; it is simply commodity fetishism. We did not win Iraq, we did not win Afghanistan, and this was due to rapid battlefield dominance during the invasion while allowing stay-behinds and saboteurs to slowly sap our morale. There was no strategy beyond initially demonstrating the omnipotence of our so-called advanced weapons. We were chasing ghosts, becoming prey, while they remained the hunters. And it appears based on the lackluster results in Ukraine that we learned nothing. Unconventional Warfare, Guerrilla Warfare, and a People's War, whatever term you prefer, appeals to that strength of being the underdog in one sense while allowing an enemy's hubris become their own undoing. In recognizing this reality signals intelligence can wreak absolute havoc without having to fire a shot simply by becoming a ghost.

    So let's say that the very same high threat group – nation state backed or no, its irrelevant – is operating in your area. They have nice equipment, night operating devices (NODs), they have clean, new weapons, decent boots and are in fighting shape. A direct fight with them may very well be suicidal, but we still must attack them. They have a radio system you recognize and have watched their electronic signature on your Spectrum Analyzer. Lots of activity in the 136-174 MHz range. Poor discipline on the radio leads to, at a minimum, an early warning on your part. In addition you now have at least two points of exploitation: their over-reliance on electronic communications and likely unfamiliarity with their equipment. You could attack them simply through passive jamming (more on this later) to degrade their communications capability while introducing doubt about the reliability of their radios. This does wonders for morale in my experience and creates that seam and gap in which the guerrilla operates. Couple that with leading them into a trap where you wound one or two in a sniper attack and the damage is done. Warfare is psychological first.

Radio Direction Finding: Creating the Tactical Advantage

Our spectrum search has established our communications baseline. Our point search has created and refined our targets, hopefully mapping the hierarchy and told us some things about their relative discipline in the process. Now its time to geolocate the source of the signal itself. This process is called Radio Direction Finding (*RDF*).

RDF begins with finding the direction, or azimuth, of an intercepted transmission. This is where the Yagi antenna described in Chapter 3 comes in handy. Once we have it oriented in the direction our signal strength is at a maximum, we now check our compass to get the azimuth. This is your magnetic azimuth. The next step is converting that magnetic azimuth to a grid azimuth and you're either adding or subtracting the declination angle for your area which is found in the map legend. You should know what the declination for your area is – its a basic map reading skill, but I digress. Once you've done your conversion, plot where you are on the map and draw a line along the azimuth. Your suspected transmitter is somewhere along that line. This is your **Line of Bearing** (LOB).

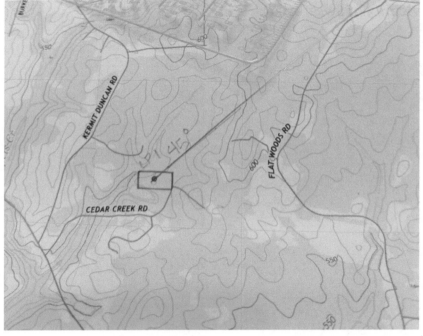

We can take this a step further. **Triangulation**, or the intersection of at least three different azimuths or LOBs, gives us a relatively confident location of the point of origin (POO) of the intercepted signal. What we need is a couple more Listening Posts (LPs) that have also intercepted the signal and drawn azimuths. Once those are shared, the point of intersection is where your hostile POO is located. Note that this is not a particularly long range, within 2 KM or so. That's well within both VHF and UHF signals range, especially from license-free equipment such as MURS or FRS. And it is also well within your distance to do something about it.

On the same map below, we've got three additional LPs spread out along that same area. Each of them intercept a signal and the corresponding LOBs are plotted on the map. The suspected POO is marked by the black X (in vicinity of Sweetbriar):

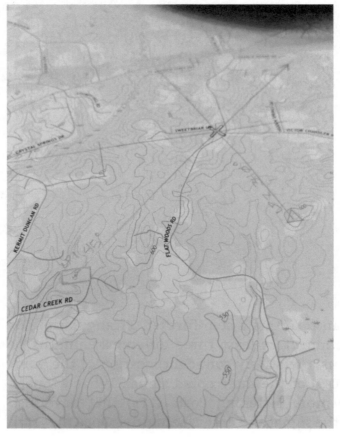

That's the manual way to do it. This is where the KrakenSDR really comes in handy and should be on your must-have list along with the Adcock Array purpose built for it from Arrow Antennas. With the built in RDF software superimposing the azimuth on a map, we can quickly gain at least one LOB which serves as our starting point. Having at least one fixed LP equipped with the Kraken makes the time on target that much faster and handed off to your other collection teams in the field.

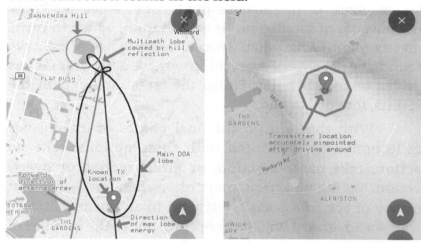

RDF is one of those topics that seems complicated, and in practice it can can be. But the basics are actually quite simple. It gives a small group of armed men direction, which in turn, grants that necessary speed and surprise when properly executed. It is a skill that requires constant practice to refine and maintain but in doing so set the tone for the next phase of signals intelligence: exploitation.

CHAPTER 5. TACTICAL SIGNALS EXPLOITATION

We've covered the Find and Fix and dabbled in the Finish phase of intelligence production. Since we know the purpose of intelligence is exploitation, which in turns means identifying what we can do with what we have found, the purpose of everything we have done thus far has driven us to this point. Finish the target and exploit the results, for good or bad.

With that said in the final phases of intelligence collection we're presented with three options: **take no action, take passive action, or direct (active or overt) action**. Taking no action, more often than not, is the best course of action especially early on in an unconventional war. An adversary will likely operate under the premise of superiority over a populace in all aspects, be it firepower, organization and certainly radio spectrum dominance. As we discussed in the last chapter this leads to hubris. Taking no action initially, or exhibiting tactical patience, allows that enemy to show his hand while concealing our own. What capabilities do they in fact have? And to that point, are they in fact competent? What patterns of life have they established for themselves that provide us later points of exploitation? It is important to stress not being over-eager to use every tool at one's disposal initially.

It has long been my suspicion that the 2008 invasion of Georgia by Russia was a probing attack in preparation of a much longer anticipated conflict. We had Special Forces teams on the ground conducting overt Foreign Internal Defense (*FID*) along with all of the technological enablers being fielded elsewhere in the world at the time – namely Iraq, Afghanistan, Yemen, and Israel (*targeting Syria and Iran*). Russia sought to know how their forces would fight against these systems and what those points of exploitation would in fact be; thus, they invaded. And while the pundits and armchair Generals may poke holes in the results, they did in fact win and in doing so gained a visceral working knowledge beyond hyperbole of the capabilities of those systems at all levels down to the small unit. This would be replicated in Ukraine.

The Russian invasion of Ukraine actually began in 2014, culminating in the overt large scale military invasion of February 2022 to the present. While their performance has been lackluster on one hand in my opinion, on the other it demonstrates a much longer term tactic remembering the success of the Soviets in rapidly taking Europe east of Berlin in the months after the conclusion of WWII. There were no more able bodied men to fight and even worse, no material or willpower to mount a meaningful resistance. It is a polar opposite of the US approach hinged upon rapid battlefield dominance which leaves a large populace still very eager to fight, albeit blending directly into the populace itself. What does this have to do with SIGINT? Instead of relying upon tactical patience, the West jumped in with both feet and no strategy, relying on Global War on Terror (GWOT) tactics of using conventional forces (*micromanaged 'special operations' troops*) in the same manner as they did on lightly armed insurgents with no electronic exploitation capabilities. Thus far it has been a complete disaster.

The best course of action the West could have taken was to allow Russia to take Kyiv and foster an insurgency in the same vein as the Baathists in Iraq circa 2004-07. They did not do this for political and financial reasons and the results, nearly two years into the conflict, have been equally lackluster despite the massive bravado to the contrary. Ukraine is left depleted of both weapons and able bodied men with no winning strategy beyond pushing the next batch of widgets to the front. In doing so, it has exposed severe weaknesses in the weapons manufacturing and procurement on part of the West. On the bright side, however, it is leading to a renaissance particularly in the field of signals intelligence and how that pertains to small unit tactics, be it nation state or non-state actors (*that'd be you, guerrilla*).

In taking no action we gain much. We do not reveal our own capabilities. Feigning weakness when you are in fact strong is a timeless virtue of Sun Tzu, after all. Persistently and relentlessly monitor, map their capabilities, hierarchies, strive to know everything about your target. Those opportunities will in fact present themselves readily when the enemy finds himself most comfortable, when he believes he has no opposition, when he believes he's won. At that point he is most vulnerable.

Jamming: Electronic Warfare

The year was 1954, the place, a small central American country called Guatemala. Jacobo Arbenz, a left-leaning populist politician is elected President and vows land reforms for the nation's poor. This damaged United Fruit Company's business model and since Arbenz, while not formally aligned with the Soviet Union, was in fact a Leftist, the decision was made in Washington that he had to go. Militarily an invasion could not be justified and Guatemalan troops left much to be desired if they could be relied upon as a complete political entity. While the senior Officers were seen as politically reliable the enlisted troops not so much so. The US thus resorted to other means conducted the first overthrow of a country strictly through Electronic and Psychological Warfare.

They began by offshore jamming of all of the radio stations with propaganda designed to scare the nation's troops into surrendering and eroding the people's confidence in the nation's leadership. They made it convincingly seem as though the counterrevolution was real despite the fact that militarily the CIA had just over a Company's worth of guerrilla fighters, somewhere in the neighborhood of 180 riflemen. But with those small numbers they rapidly took an armed outpost and convinced the abdication of the Arbenz government. Electronic warfare was the real hero of the battle, in fact feigning strength from a position of numerical weakness and in doing so writing a new doctrine on psychological and electronic warfare.

In 2021 Cuba faced a populist uprising in a major challenge to the communist government of Miguel Canal-Diaz. The old face of the revolutionary party, the Castro brothers, had faded into obscurity and the reality of collective poverty and the permanent status of poverty under the communist regime led to riots in all of the major cities of Cuba. Decades of government paranoia and repression left them with few options for communicating their tactics in organization in real time. The possession of two way radio is tightly controlled and where it is allowed it is approved for government propaganda alone. That said there were a plethora of High Frequency (*HF*) radios capable of regional and global communications and, due to the level of poverty in the island nation, most of the radios for entertainment are shortwave AM (*which operate in the HF frequency range*). The Radio Recon Group took to the airwaves in the amateur radio bands in AM (*not Single Sideband, or SSB*) to transmit words of encouragement and news from the outside world to them. We were met with jamming from what looked similar to an over-the-horizon radar system. Several of our operators obtained lines of bearing on the source and it led, not surprisingly, to Cuba. The central government had taken its radio stations off the air (*Radio Havana and Radio Rebelde*) in an effort to control and isolate the source of the RF emissions inside the country while targeting our signals. While they were successful in putting down the uprising, it serves as another example of how electronic warfare is implemented in a Guerrilla War.

There are two broad courses of action through electronic warfare: Passive and Active Jamming. Jamming a signal, in both forms, requires targeting the source of that signal with an equal signal stronger than the one being emitted. Picture yourself once more in that dark room. You're shining a light which is your signal that is being emitted. All of a sudden a strong light shines, blinding you and drowning out whatever energy you were emitting. If you can picture this, then you have a visual representation of the process of jamming.

Passive Jamming is the simple introduction of energy during an enemy's transmission. There is no other noise or unique signal being transmitted. The purpose behind this is to degrade the target's confidence in his equipment. Passive jamming also utilizes a technique known as Electronic Isolation of a Target. Our goal is to selectively attack targets of interest, not shut down an entire unit at once. We want to sap their confidence in themselves and their equipment and that comes with targeting the mid level leadership once they've been identified.

For example, you're observing a Company of occupation troops in an area and have mapped their communications capabilities. Our target is the mid-level leadership. That consists of the squad leaders and the respective platoon sergeants. You know, the guys who're actually running things. Once you identify those callsigns those become your targets. When they go to transmit, you'll transmit too. Only when you do it you're aiming that Yagi antenna you built in their direction, drowning their signal out just as with the example above. Yagis are incredibly effective weapons on the electronic battlefield not just for reception but for transmitting.

But we're not done. The jamming is not constant; we have to assess the results. Maybe they replace a radio. Maybe they change frequencies. Maybe they ignore you. But in any case we have to constantly assess what changes our target made, if any, in order to continue our electronic attack. Another technique is a play on the complex or baited ambush, leading a target into a trap. Remember, you're here to make his life a living hell in its final moments. Conducting a sniper ambush, then jamming their signal when they're coordinating the response to cause even further chaos would devastate morale. At this point we would want to look beyond the mayhem we've caused to analyze the intended effect – what changed about our target? Did they change practices or operating principles? Did they abandon otherwise useful equipment due to our jamming? These are all of the analytical questions we have to follow up on in order to further refine our techniques.

Active Jamming is the introduction of false traffic or psychological warfare to the operating frequencies of the target. This is where traffic is played back over the radio in order to induce confusion or fear. At this point it is likely known to the target if they have any awareness at all that they're being jammed, but active jamming is utilized to create confusion or distrust in the traffic itself.

An example is the aforementioned Armored Brigade that was being jammed and was, at least initially, unaware. The unit Commander had a bad habit which had been reinforced by GWOT thinking and the fact that we had no sophisticated electronic warfare opponent. He would get on the radio and give the unit briefing to the subordinate Commanders every morning at a specific time. This established a distinct pattern of life which in turn also granted PID of not just his callsign but his voice. Once the electronic opposing force had mapped the SOI they recorded his voice and instructions, playing it back the following morning and inducing quite a few problems. Once the unit realized they were being jammed they implemented a COMSEC changeover which was in turn found out and jammed again in short order. It was not the technology that was the problem, it was his technique.

Another example was the Special Forces communications of the MAC-V in Vietnam. The North Vietnamese Army had aid from the Soviet Signals Intelligence groups, and very likely aid from the anti-war groups in the United States, to play recordings of family members' names and personal information of the troops over their radios as a psychological warfare tool. On the

American side Operation Wandering Soul, or recordings of villagers giving their sorrowful accounts of dead Viet Cong, were frequently played over loudspeakers and radios in areas with high insurgent activity.

Assessing the Results

The purpose of jamming, be it passive or active, is to introduce doubt in the minds of your target, whoever they may be. That doubt manifests itself first in the form of a pause. That pause may in fact be all a guerrilla is after – the reactionary gap – in which we operate. In order to know that gap we have to take a step back and analyze what was achieved, for both good and bad. For every action there is an equal and opposite reaction, and that outcome is not universally favorable. There exists a delicate balance much like a poker game. Don't play your cards too soon but don't overvalue your hand, either. We must know what was achieved while recognizing what we may have lost.

Another example of electronic warfare, this time gone wrong. In Afghanistan we implemented a plan of destroying Taliban repeaters. The SIGINT teams had located and mapped most of them and the scheme of maneuver was to simply blow those repeaters up. Great, now the Taliban can't communicate. GWOT is won, right? Well, all we actually did was blind ourselves, at least in the near term. The Taliban previously had relied on those repeaters to coordinate their movements of supplies from Pakistan and previously had given no indication that they were aware of our eavesdropping. Once we took kinetic action, however, the SIGINT teams were now back to square one. The Taliban resorted to communicating peer to peer and changing their frequencies at uneven intervals which in turn deafened what had been our greatest attribute. We acted too early and in our assessment it was realized that acting too aggressive, too soon, became a major detriment to capabilities. It is my personal belief that we would have had a much higher degree of success early on, but, I digress. We have to take the good and the bad in an effort to refine future practices and that requires a strong degree of humility.

Signals Intelligence is the Weapon

All of the rifles and the men who wield them are for naught without purpose and direction. Signals intelligence, as a byproduct of the proliferation of newer, faster and more seemingly omnipotent methods of communications partnered with the ever-increasing desire for micromanagement, has become the primary means by which that intelligence is fomented. Fortunately that skillset and the equipment necessary is rapidly becoming less and less expensive to the common man. It is my personal belief that communications receivers, software defined radio, and the evermore impressive technology driving that revolution is every bit as important to a free people as a rifle ever will be, perhaps even more so. It is the tool by which the other weapons are aimed.

While it is disheartening on one hand to see the topic so frequently misunderstood and thus overlooked altogether, this work's aim has sought to at least create a primer by which to build. Like anything, this skill is not a one and done but rather a continuum of training and building knowledge. I have said in the past that technology remains in perpetuity while techniques are timeless. The technology will change through natural evolution, but those underlying skills of how we hunt will not. We do in fact kill people based on metadata. And with that innate desire among those forces of oppression to seek greater levels of micromanagement thus opens the door to further and further points of exploitation. For every move there is a counter. It is the duty of a free people to understand those moves in order to remain so.

- NC Scout, 2023

APPENDIX A: EXPLOITING INDIVIDUAL SOURCES OF COMMUNICATIONS

License Free Communications Bands:

The first target of interest is the license free frequency ranges in the US. These are relatively easy to exploit, with the exception of Citizen's Band (CB) due to the sheer volume of traffic and the propagation characteristics of the frequency range. These will likely be used by an unsophisticated target, but not in all cases. Analog signals found in this range are easily jammed.

I load these into the memory of a Baofeng, with the exception of CB *(its outside the frequency range)* and set it to scan along with monitoring on the Spectrum Analyzer.

License Free Frequencies:
- **Citizen's Band (CB)**
  - **CB 1:** 26.965 **CB11:** 27.085 **CB21:** 27.215 **CB31:** 27.315
  - **CB 2:** 26.975 **CB12:** 27.105 **CB22:** 27.225 **CB32:** 27.325
  - **CB 3:** 26.985 **CB13:** 27.115 **CB23:** 27.235 **CB33:** 27.335
  - **CB 4:** 27.005 **CB14:** 27.125 **CB24:** 27.255 **CB34:** 27.345
  - **CB 5:** 27.015 **CB15:** 27.135 **CB25:** 27.245

- **CB35:** 27.355
  - **CB 6:** 27.025 **CB16:** 27.155 **CB26:** 27.265
  - **CB36:** 27.365
  - **CB 7:** 27.035 **CB17:** 27.165 **CB27:** 27.275
  - **CB37:** 27.375
  - **CB 8:** 27.055 **CB18:** 27.175 **CB28:** 27.285
  - **CB38:** 27.385
  - **CB 9:** 27.065 **CB19:** 27.185 **CB29:** 27.295
  - **CB39:** 27.395
  - **CB10:** 27.075 **CB20:** 27.205 **CB30:** 27.305
  - **CB40:** 27.405

- **Multi-Use Radio Service (MURS)**
  - **MURS 1:** 151.820
  - **MURS 2:** 151.880
  - **MURS 3:** 151.940
  - **MURS 4:** 154.570
  - **MURS 5:** 154.600

- **Family Radio Service (FRS) & General Mobile Radio Service (GMRS)**
  - **1:** 462.5625   **13:** 467.6875
  - **2:** 462.5875   **14:** 467.7125
  - **3:** 462.6125   **15:** 462.5500
  - **4:** 462.6375   **16:** 462.5750
  - **5:** 462.6625   **17:** 462.6000
  - **6:** 462.6875   **18:** 462.6250
  - **7:** 462.7125   **19:** 462.6500
  - **8:** 467.5625   **20:** 462.6750
  - **9:** 467.5875   **21:** 462.7000
  - **10:** 467.6125   **22:** 462.7250
  - **11:** 467.6375
  - **12:** 467.6625

Amateur Radio (VHF / UHF)

Amateur Radio, or "ham" as its nicknamed, carries the metadata of its licensed users through the callsign which is assigned by a nation's governing entity.

The frequency allocations themselves are important for understanding where this traffic will be found. For SIGINT purposes it is important to note that in an asymmetric conflict a guerrilla force will very likely be using repurposed ham radio equipment.

US Amateur Radio (HAM) Frequency Allocations
- **160m:** 1.8-2mHz
- **80m:** 3.5-4mHz
- **60m:** (1)5.332mHz (2)5.348 (3)5.3585 (4)5.373 (5)5.405
- **40m:** 7-7.3mHz
- **30m:** 10.1-10.150mHz CW/DATA ONLY
- **20m:** 14-14.350mHz
- **17m:** 18.068-18.168mHz
- **15m:** 21-21.450mHz
- **12m:** 24.890-24.990mHz
- **10m:** 28-29.7mHz
- **6m:** 50-54mHz
- **2m:** 144-148mHz
- **1.25m:** 222-225mHz
- **70cm:** 420-450mHz
- **33cm:** 902-928mHz
- **23cm:** 1270-1295mHz

Digital Mobile Radio (DMR)

DMR is one of the more common digital protocols found in handheld radios. It offers decent build quality in most of the handhelds, the ability to send SMS messages and native *(built in)* encryption of up to AES-256 in the higher end models. DMR operates in Time Division Multipath Access (TDMA), meaning it assigns each signal an individual time slot based on an internal clock. In many models that clock is maintained by a passive GPS, GLONASS, or BeiDu signal. It is a good choice for inter-team communications, however, it is not perfect.

Exploiting DMR can be done in a couple of ways. In a

tactical or time sensitive environment, you're not likely to break the voice encryption being used. This is true for most forms of encryption, even the weaker ones. Bypass that altogether. What is being transmitted in addition to the traffic is the radio ID, the color code and time slot. These are the message headers, or, metadata. You can find these using DSD+ and an SDR. This gives us positive ID of our targets. Considering DMR is most often utilized in a tactical role, RDFing the source of the signal and closing with the target is the best solution.

While DMR is jam resistant due to the narrow bandwidth of the mode and the exchange of metadata to ignore other types of signals on the same frequency. The weakness is the equipment itself. Handheld radios are not built to withstand high amounts of directed energy on a nearby frequency. This is where your Yagi comes in handy once you know the LOB to the target. You're not jamming the signal, you're overloading the radio itself and potentially permanently destroying it by deafening the receiver.

Most Common Frequency Range: 136-174 MHz, 400-470 MHz

DMR Waterfall Image:

P25

Project 25 (P25) is a digital voice and data mode in common use with public safety agencies. It is encryption capable and is similar in operation to DMR. P25 is the most common mode for Law Enforcement in the United States. It does however share all of the same pitfalls of DMR especially in transmitting metadata. The system was also explicitly designed to be used in conjunction with repeaters and its equipment reflects this. While some of these radios show up on the surplus market, they are highly exploitable due to their limited frequency range and easily identifiable mode.

The best method of exploitation for P25 is RDF of the source signal followed by jamming that source. The equipment itself is built to handle relatively low amounts of power (5-10w) and the directed energy from a Yagi on its operating frequency will likely burn the receiver quickly.

Most Common Frequency Range: 136-174 MHz, 380 MHz, 512 MHz, 769 MHz, 824 MHz, 851 MHz, 859 MHz

Waterfall Image:

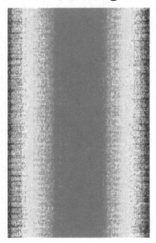

Commercial Drones

Without a doubt drones have made the biggest impact on warfare tactics since the advent of the machine gun. As a result much has been made of their detection and early warning. They are a weapon system like any other and do, in fact, have certain major drawbacks. Commercial drones especially so. The war in Ukraine has taught many painful lessons in electronic warfare on both sides, many of which have been covered in this book, but especially so concerning the proliferation of commercial drones.

Drones emit several signals simultaneously – most notably ADS-B, which can be detected on both an SDR system as well as the handheld PortaPack H2. The KarkenSDR works extremely well in this role and can be used as a fixed passive radar system for detecting and monitoring their presence. That aside, there are two frequency ranges, both in the WiFi range they operate on as a control frequency for the pilot and the signal for the data being returned to the remote control. This is assuming they're not using a cell phone for that control, which is another point of exploitation we'll be covering in and of itself. The remote is emitting a signal as is the bird.

It must also be noted that drones are built with highly sensitive receivers designed to handle very low amounts of power. As with DMR and P25, near field energy can quickly destroy the receiver in the drone, likely dropping it from the sky. This is why we're seeing the proliferation of jammer guns, which are just transmitters in the WiFi range with a high-gain Yagi that can be aimed and fired from the shoulder.

Frequency Range: 2.400–2.483 GHz and 5.725–5.825 GHz

Waterfall Image:

NATO Military Ground Communications (SINCGARS)

The Single Channel Ground / Airborne Radio System, or SINCGARS, has been the NATO standard for ground communications since the late 80s/early 90s. In addition to operating in a single channel mode (one frequency), it also has the ability to frequency hop. This is done every three seconds and relies on the internal clock to sync the hops. It can be loaded with NSA Type III encryption. SINCGARS, in both the AN/PRC-119 and 117G currently in use as mounted or backpack radios, along with the PRC-152 as a handheld, operates in the 30-88 MHz range for ground communications and is relatively easy to detect. Operating in the lower end of VHF it has a great line of sight range for reliability of communications but this makes it easy to detect and in turn RDF the source of the signal.

While it can be jammed with off the shelf equipment in theory, the best course of action against SINCGARS is finding a LOB to the source of the equipment itself as an early warning or geolocation of your enemy for mission planning purposes.

Frequency Range: 30-88 MHz Ground, 250-315 MHz SATCOM

Russian / CSTO AZART Radio System

The latest radio system to be fielded by the Russian military is the R-187-P1E AZART. This represents the latest counterpart to the SINCGARS system with the inclusion of Tactical Satellite (TACSAT) communications capability. In utilizes a modified version of the TETRA digital protocol for its built in communications security and encryption capability. Several of the first generation have been captured in Ukraine and users have complained of short battery life.

Like SINCGARS, the best course of action with AZART is identifying the LOB to the point of origin for targeting purposes. While again, in theory, the COMSEC on the radio could be broken in real time, this is unlikely as any potential exploits discovered now are likely to be reverse engineered and prevented in future conflicts.

Frequency Range: 27-530 MHz

Waterfall Image:

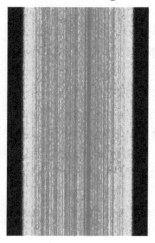

Cell Phones

Far and away the most common devices in any working environment are cell phones. That said its also the most highly exploitable item people carry on themselves. Much is made from the internet networking and data sharing end of phones and there certainly is a case for that which can be resolved by using one of the more privacy-focused operating systems such as CalyxOS or GraphineOS on an unlocked device. This is only part of the problem, however. The heart of the issue is the way cell phones themselves function and access a network.

Cell phones have at a minimum three pieces of data assigned to them: the **International Mobile Equipment Identifier** (IMEI), the **International Mobile Subscriber Identification** (IMSI) and the network or **mobile phone number** itself. The IMEI is a serial number assigned to your device. All devices have them and they continuously transmit that data. The IMSI number is assigned to the SIM card and is also transmitted constantly along with the IMEI. Your carrier number is what people dial to access your device. The IMEI and IMSI are both registered with the network at the time of the phone's activation. They are never shut off and this data is accessible to anyone within range of receiving it. Even if one pulls the SIM card from a device the IMEI is still transmitting from what's known as the baseband of the phone. Again, there is no way to shut this off short of desoldering the antenna from the device, but this is akin to transmitting with your radio with no antenna – it'll burn itself up. Without a SIM card any phone in the US can still call 911. That should tell you at a minimum its still transmitting and thus can be located (*and rather easily*).

To make matters worse cell phones are constantly being triangulated by the cell phone towers themselves regardless if they're your carrier or not. This is how they are designed to function and they are constantly logging and reporting that data to the phone companies. Internationally this is how all cell phone networks function – they serve ostensibly as a means of communication for a populace but not without a high level of governmental monitoring and control. It may sound jarring to some but the US has been doing this long before the Patriot Act.

All of this being said, that IMEI and IMSI number is a problem and the lone way to mitigate it is simply abandoning cell phones if security is the goal. Keeping that in mind, there's a number of ways civilians can exploit cell phones. The first is through simply jamming the signal itself. This can be done with the PortaPack H2, BladeRF and HackRF to name a few. Since cell phone signals in all of the bands across the world are designed to operate with relatively low power outputs at the higher end of UHF, those signals are easily interrupted by a closer, more powerful source.

But perhaps we don't want to jam, we simply want to collect. The government has been employing mobile data skimmers, the most famous one called the Stingray, for a long time. We used these to high effect in Afghanistan and Iraq and it remains the primary means by which we target high value individuals (HVI). This was specifically what General Hayden was referencing when he made his statement regarding the killing of people based on metadata. That metadata is the IMSI, IMEI and geolocation.

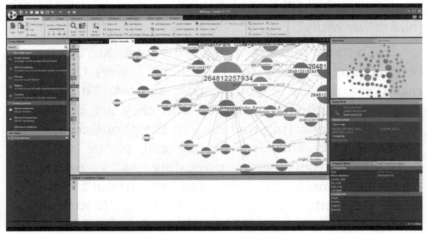

You can build your own IMEI & IMSI skimmer with the same aforementioned tools. DragonOS comes with an IMSI logger built in as well. GR-GSM is one of them. From here, once we've created a log of IMEI and IMSI numbers active within our are of reception, I can track the geolocation and patterns of those numbers with an open source data scraping program known as Maltego. From here I can analyze not only the target's behaviors, but who he's linked to as well giving me new targets of interest.

This is what's available on the open source side. Any governmental tool kit already has every phone mapped out and can track them in real time – thus, anything relying upon cell phones for their operation should be written off for a potential guerrilla force. Any theoretical benefit is negated by the fact that the phone itself is not your friend nor is any convenience you may imagine it provides.

Mesh Networking: LORA, GoTenna and ATAK

Mesh networking in a nutshell is the linking of several small, low powered transmitters to create an ad-hoc data network. These have been promoted as a plug n' play option for creating communications in austere environments originally with the intention of disaster relief. These provide texting capability. For that role, its still mediocre at best and utilized by those seemingly incapable of land navigation, radio communication, or any other skill requiring prior competence. It appeals to both micromanagers and those who cannot fathom existing without being wired into a device. That it appeals to any military unit in the face of a peered adversary is alarming given what we now have learned from Ukraine.

The real problem arises when its utilized in a tactical role. Mesh networks require a consistent link in order to function and work like a beacon. Anyone can intercept this regardless of encoding of the contents of the message. These operate in the 902-928 MHz range and are extremely easy to spot, even in a high electronic noise environment, due to their unique transmitter signature. Their Mobile Ad-hoc Network (MANET) counterparts operte in the same 2.4-5.8 GHz range as WiFi creating a unique signature as well. In addition, since they're essentially beacons, they are easy to isolate and obtain a LOB to the suspected source utilizing KrakenSDR.

Each of these are predicated upon their use with android phones. Since we've already covered why this is an incredibly bad idea, it should be self evident how to target and exploit these signals. ATAK in particular transmits your geolocation based on the phone's geolocation. This is in turn triangulated by the cell phone towers. Bad idea.

Exploiting these devices is quite simple. After geolocating the source of the signal itself, you'll likely be able to pair that data to their phone data you've also collected and map their patterns of life. But even if we're not going that far, closing with and giving them their worst day is certainly an option as well. If they are lazy enough to be using these devices in a tactical environment they're likely easy prey who've taken the easy road on everything else. Once upon a time capturing a target's cell phone was a jackpot, but then again, those were dusty cavemen. But they won.

APPENDIX B: SIGNALS INTELLIGENCE TASKING CHART (F3EAD)

- **FIND:** MAP RF / ESTABLISH SITUATIONAL AWARENESS / FIND ELECTRONIC TARGETS
  - Task: Determine signals baseline for an area
    - **Spectrum Search:** Map all detected RF
      - Tools:
        - SDR (KrackenSDR, RTL-SDR, BladeRF, etc)
        - Digital Scanner (Uniden SDS100)
        - Spectrum Analyzer (TinySA Ultra)
        - Communications Receiver (AOR DV10)
          - Antenna: Omni Directional
    - **Point Search:** Determine purpose / frequencies of strongest detected signals
      - Tools:
        - SDR
        - Digital Scanner
        - Communications Receiver
          - Antenna: Omni Directional
- **FIX:** ESTABLISH LINES OF BEARING TO TARGET / MAP PATTERNS OF LIFE
  - Task: Begin Electronic surveillance / Determine LOB to target

- Tools:
  - Recorder
  - SDR (in fixed positions only)
  - Communications Receiver
  - Spectrum Analyzer
- Antenna: Yagi. Adcock
- Task: Determine SOI of targeted signals
  - Who is in charge?
  - What is this unit's purpose?
  - What is their overall morale?
  - What is their level of experience?
  - What is the potential outcome of exploitation?
  - **FINISH:** Initiate Electronic Attack
- **Passive Jamming:** Introduce doubt in equipment or skill level
  - Tools:
    - HackRF
    - Portapack H2
    - Transmitter for intended frequency range
  - Antenna: Yagi
- **Active Jamming:** Psychological Warfare on Target
  - Replay previous transmissions
  - Transmit demoralizing recordings
  - Tools:
    - Transmitter for intended frequency range
- **EXPLOIT:**
- Identify Reactionary Gap
- Identify additional Behavioral Changes

- Identify Action Indicators
  - **ASSESS:**
- What was overall effect?
- How did target change practices?
- What did we gain?
- What are our new targets?
  - **DISSEMINATE:**
- Best practices for future targets

Made in United States
Troutdale, OR
12/21/2024